Mathematics Tests for Selective and Independent Schools and for Scholarship Examinations

* 20 practice papers
* all multiple choice questions
* based on previous papers
* answers to all questions
* outline of solutions to nearly all questions
* ideal for entrance into New South Wales and Victorian Selective Schools
* ideal for entrance to Independent Schools throughout Australia
* ideal for Scholarship and Bursary Examinations

by James An, Jim Coroneos and John Smith

Item 49 - New Revised Edition

First published 1991
Reprinted 1992, 1993, 1994, 1995, 1996, 1998
New Revised Edition 2000

This book is available from any recognised bookseller or by contacting

James An
Academic Education College
77 Cowper Street
CAMPSIE NSW 2194
Ph: (02) 9789 2645
Fax: (02) 9718 1011

Jim Coroneos
Coroneos Publishing Co. Pty. Ltd.
PO Box 25
ROSE BAY NSW 2029
Ph: (02) 9371 8530
Fax: (02) 9371 7099

John Smith
Smith Mathematics Coaching
14 Kooloona Crescent
PYMBLE NSW 2073
Ph: (02) 9498 8883
Fax: (02) 9498 4118

Typesetting: TDC
Todays Digital Communications
PO Box 1047 Darlinghurst 1300
Phone: (02) 9261 0550
FAX: (02) 9268 0590

2024 10 04

CONTENTS AND SCORE

About this book

This book is designed to help students excel at three mathematics examinations:

1 The Year 6 Selective Schools Examinations for entrance to NSW and Victorian Selective State High Schools, and

2 The Year 6 Entrance Examinations for Independent Schools

3 The Year 6 Scholarship and Bursary Examinations for Independent Schools

The first twelve papers are practice examinations for entrance to Selective and Independent Schools; the last eight papers for the Scholarships and Bursaries Examinations.

Students and teachers may also use these tests for:

4 Independent Schools entrance examinations to any year, and

5 General mathematical ability

How to use this book

No matter which examination is being attempted, the student should sit for each of our papers. The format of each paper is based on previous papers and the questions have been carefully chosen to mirror the style and difficulty of the actual examination.

We have adhered to the format used in previous papers. Questions may be more demanding than those that you are used to and you should strive to understand the techniques involved in their solution. Do this diligently and you will go in to your examination knowing that you will do well.

The recommended time for each of these papers is 40 minutes. However, you should complete each paper regardless of the time limit.

Good luck in your studies!

Selective Schools
Paper 1

1	The average of six numbers is 4. A seventh number is added to the first six. The average of the seven numbers is 5. The seventh number is	(A) 4½ (C) 2	(B) 7 (D) 11
2	How many years are there in 2 centuries and 2 millenia?	(A) 220 (C) 2 200	(B) 202 (D) 2 020
3	How many times can 2½ be subtracted from 12½?	(A) 5 (C) 7	(B) 6 (D) 4
4	1.21 - 0.9 =	(A) 1.12 (C) 0.31	(B) 0.12 (D) 1.31
5	What is the lowest common multiple of 3,4 and 6?	(A) 6 (C) 4	(B) 12 (D) 24
6	Which of the given numbers has the largest value?	(A) 1.2 (C) 1¼	(B) 1.24 (D) 1.19
7	Simplify 1½ + 2⅖	(A) 3⁹⁄₁₀ (C) 3½	(B) 3⅔ (D) 3⁷⁄₁₀
8	When $299.99 is rounded to the nearest dollar it becomes:	(A) $299.00 (C) $299.90	(B) $290.00 (D) $300.00
9	John saw frost on the grass in the morning. What was the likely temperature?	(A) -10°C (C) 15°C	(B) 1°C (D) 30°C
10	If 120 is three tenths of a number, the number is:	(A) 240 (C) 400	(B) 120 (D) 150
11	How many whole hundreds are there in 2 967 890?	(A) 2 967 (C) 29 670	(B) 29 678 (D) 8
12	What is the average of 3.1, 8.1, 5.1, 7.1, 3.1 and 4.1?	(A) 7.1 (C) 5.6	(B) 6.1 (D) 5.1

13 A rectangular fish tank is 50cm long, 40cm wide and 30cm deep. How many litres of of water will it hold?	(A) 60 (B) 6 (C) 30 (D) 3
14 The place value of 7 is 219.07 is	(A) Hundreds (B) Hundredths (C) Units (D) Tenths
15 What number is represented by the □? 0 — $1\frac{1}{5}$, $1\frac{1}{5}$, $1\frac{1}{5}$, $1\frac{1}{5}$, $1\frac{1}{5}$ — □	(A) 5 (B) 6 (C) $5\frac{1}{2}$ (D) None of these
16 A man wishes to join the police force. He weighs 100kg and must lose 200g each week in order to be accepted. What would he weigh 50 weeks from now?	(A) 80kg (B) 85kg (C) 90kg (D) 95kg
17 24kg of flour cost $28.80. How much can be purchased for $3.60?	(A) 1kg (B) $1\frac{1}{2}$kg (C) 2kg (D) 3kg
18 Eight million eight hundred and eight is equal to	(A) 8 800 008 (B) 8 080 080 (C) 8 000 808 (D) 8 008 008
19 Which is equal to half a quadrant of a circle? (O is the centre of the circle and sketches are not to scale.) (A) 0 (B) 0 (C) 0 45° (D) 30° 0	(A) (B) (C) (D)
20 In the expression 27 + 36 ÷ 4 x 7, which is calculated first?	(A) 27 + 36 (B) 36 ÷ 4 (C) 4 x 7 (D) 36 x 7
21 Two painters were contracted to paint a building for $2 265. The cost of the paint was $465. If they took 6 days to do the work, how much per day did they each make?	(A) $300 (B) $250 (C) $200 (D) $150

22	What is the most probable missing number in the pattern: 3, 7, □, 15, 19?	(A) 9 (B) 10 (C) 11 (D) 13
23	Add the product of 28 and 3 to the sum of 29 and 11. The answer is:	(A) 124 (B) 104 (C) 1 204 (D) 907
24	26 marks out of 50 expressed as a percentage is:	(A) 26% (B) 52% (C) 50% (D) 51%
25	The number 21.87 is the same as:	(A) 2 + 1 + 8 + 7 (B) 20 + 1 + 8 + $\frac{7}{10}$ (C) 20 + 1 + $\frac{8}{10}$ + $\frac{7}{100}$ (D) 2 + 1 + $\frac{8}{10}$ + $\frac{7}{100}$
26	Which set is arranged in descending order?	(A) {3.99, 4.01, 4.10, 4.11} (B) {4.11, 4.01, 4.10, 3.99} (C) {4.11, 4.10, 4.01, 3.99} (D) {4.11, 4.10, 3.99, 4.01}
27	Of the angles 70°, 130°, 60°, 20°, 90°, 180°, 150°, which represents the acute angles?	(A) 70°, 60°, 90° (B) 20°, 60°, 70° (C) 70°, 130°, 180° (D) 130°, 180°, 150°
28	There are four shapes. Which shape does not belong to this group? (A) (B) (C) (D)	(A) (B) (C) (D)
29	Which is true?	(A) 4 716 ÷ 1 000 = 47.160 (B) 4 716 x 1 000 = 471 600 (C) 4 716 + 1 000 = 5 716 (D) 4 716 - 1 000 = 4 616
30	A bus travelled from Melbourne to Sydney in 12 hours 59 minutes. If it left Melbourne at 11:35 a.m. on Monday, when did it arrive in Sydney?	(A) 1:24 a.m. Monday (B) 1:24 a.m. Tuesday (C) 12:34 a.m. Tuesday (D) 12:34 a.m. Monday
31	At a school of 100 pupils, 30 are boys. What percentage of the pupils enrolled are girls?	(A) 50% (B) 60% (C) 70% (D) 30%

	Question	Answers
32	If the two end digits of 4186 were interchanged, the number would be made:	(A) smaller by 1 990 (B) larger by 1 998 (C) smaller by 1 998 (D) larger by 1 990
33	A 125mL carton of milk costs 20 cents. How much per litre would that be?	(A) \$1.60 (B) \$2.00 (C) \$2.20 (D) \$2.40
34	A parallelogram is a quadrilateral with …… pair(s) of opposite sides parallel. What is the missing number?	(A) 1 (B) 2 (C) 3 (D) 4
35	Instead of subtracting a number from 784, Michelle added and got a total of 1 093. What was the correct answer?	(A) 299 (B) 475 (C) 784 (D) 1 093
36	\$6 432.25 + \$2.71 + \$91.24 + \$241.76 + \$7.99. The answer is:	(A) \$6 577.95 (B) \$6 757.59 (C) \$6 775.95 (D) \$6 775.59
37	Which fraction will fill the gap in the series $\frac{4}{5}$, $\frac{39}{50}$, $\frac{19}{25}$, $\frac{74}{100}$, □, $\frac{7}{10}$?	(A) $\frac{18}{25}$ (B) $\frac{3}{5}$ (C) $\frac{79}{100}$ (D) $\frac{17}{24}$
38	These are two identical circles. Points B and D are the centres of the circles. If AB is 4cm, then BE is: 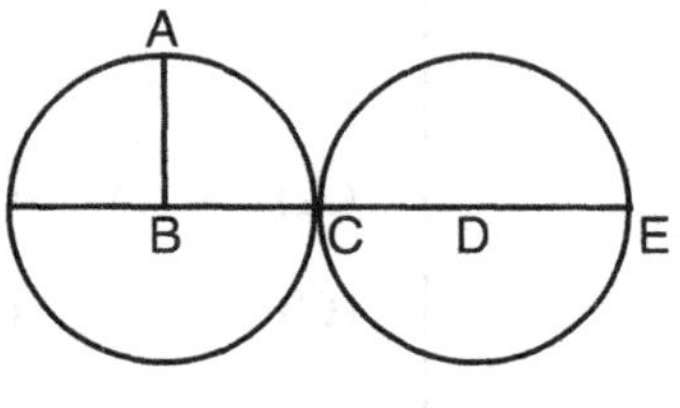	(A) 4cm (B) 12cm (C) 8cm (D) 6cm

Questions 39 and 40 refer to this symmetrical diagram - units are in centimetres.

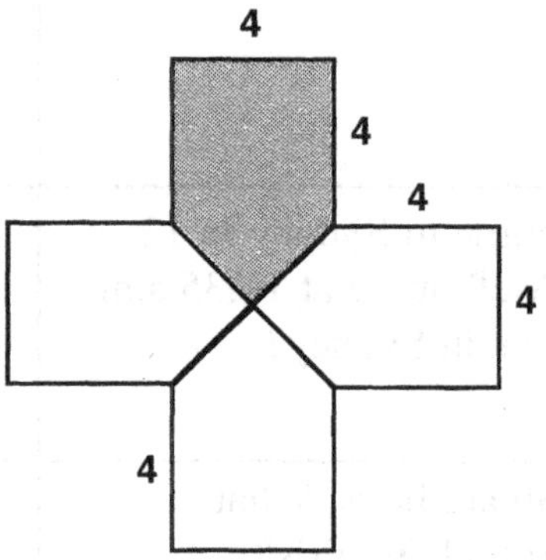

39	What is the perimeter of the above shape?	(A) 40cm (B) 44cm (C) 48cm (D) 52cm
40	What is area of the shaded region?	(A) $16cm^2$ (B) $20cm^2$ (C) $24cm^2$ (D) $32cm^2$
41	9 litres of water are to be poured into a tub using a jug which holds ¼ litre of water. How many jugfuls are necessary to pour the water completely into the tub?	(A) 2¼ (B) 24 (C) 36 (D) 48
42	If $5 in Australian money equals $16 in Hong Kong money, how much Hong Kong money should I get if I change $64 Australian for Hong Kong money?	(A) $20 (B) $90 (C) $320 (D) $204.80
43	What is the sum of the number of edges and vertices on a rectangular prism?	(A) 6 (B) 12 (C) 16 (D) 20
44	Write the third largest number you can using 4, 9, 6, 8, 0, 2.	(A) 986 240 (B) 986 204 (C) 986 420 (D) 986 402
45	An agent sells a block of land for $120 000. She receives 5 cents in the dollar commission on the first $40 000 and 2 cents in the dollar on the remainder. In total she receives	(A) $2 000 (B) $1 600 (C) $3 600 (D) $8 400

Selective Schools
Paper 2

1	The sum of two prime numbers is 28 and their difference is 6. What are the two numbers?	(A) 12, 16 (B) 19, 9 (C) 11, 17 (D) 13, 15
2	Write the second largest number you can using the digits 2, 7, 1, 5, 6 once only.	(A) 76 512 (B) 76 251 (C) 76 521 (D) 76 125
3	Which number sentence is true?	(A) 2 ÷ 1 = ½ (B) 2 x 1 = 1 (C) 2 + 0 = 2 (D) 2 - 0 = 0
4	298 is divided by 4. Which of the given possibilities is not the correct answer?	(A) 74½ (B) 74¼ (C) 74.5 (D) 74 remainder 2
5	Which of these expressions is equal to 279 x 34?	(A) (279 x 30) + (270 x 4) (B) (279 x 30) + (200 x 4) (C) (279 x 30) + (279 x 40) (D) (279 x 30) + (279 x 4)
6	Write a decimal equivalent to 11 wholes and 49 thousandths.	(A) 11.49 (B) 11.409 (C) 11.049 (D) 1.049
7	Which gives the smallest number?	(A) 6.24 ÷ 3 (B) 6.24 x 3 (C) 6.24 - 3 (D) 6.24 + 3
8	Arrange 0.6, 0.56, 0.65, 0.61 in order from the smallest to the largest.	(A) 0.56, 0.65, 0.6, 0.61 (B) 0.56, 0.6, 0.65, 0.61 (C) 0.56, 0.6, 0.61, 0.65 (D) 0.6, 0.56, 0.61, 0.65
9	A phone call to England costs $7.50 for the first 3 minutes and $2.50 for each minute after that. If a phone call to England costs $40, how many minutes did the call last?	(A) 16 (B) 13 (C) 8 (D) 14
10	The temperature in Sydney was 6 Celsius degrees above average, just before the southerly hit. After it came, the temperature fell so that it was then 7 Celsius degrees below average. What was the difference in temperature in degrees Celsius?	(A) 1 (B) 6 (C) 7 (D) 13

11	Company A can produce 10 computers per month while Company B can produce 291 computers over 3 years. At the end of one year, how many computers would you expect the companies to have produced?	(A) 97 (B) 197 (C) 207 (D) 217
12	308 ÷ 14 = 22. To find 308 ÷ 7, I should:	(A) halve 14 (B) double 22 (C) halve 22 (D) triple 7
13	I left home with \$30 and spent ¼ of my money at the grocers and ⅓ at the bakers. How much did I have left?	(A) \$17.50 (B) \$10 (C) \$12.50 (D) \$20
14	How many years are there in 20 centuries and 11 decades?	(A) 211 (B) 2 011 (C) 2 110 (D) 2 101
15	A book has 27 lines on each page. On which page will the 998th line appear?	(A) 35 (B) 36 (C) 37 (D) 38
16	Which lines are parallel? 1 2 3 4 5 6	(A) 1 and 4 (B) 2 and 3 (C) 1 and 5 (D) none of these
17	1, $\frac{1}{8}$, $\frac{1}{27}$, $\frac{1}{64}$, $\frac{1}{?}$, … In the above series, what is the most likely missing number?	(A) 125 (B) 101 (C) 83 (D) 91
18	What is the area of the rectangle in ha? 3km 1km	(A) 3km² (B) 3ha (C) 30ha (D) 300ha
19	A, B, C, D; 26.5km, 38.4km, 27km, 35km The sketch shows four towns A, B, C, D. The distances between the towns are shown. Ken lives at A and travels to C via B. He then returns to A via D, each working day (5 days) in a week. How far does he travel in one week?	(A) 126.9km (B) 1 269km (C) 634.5km (D) 649km

20	Carol's dad can purchase a video recorder for \$650 cash or he can pay it in 52 weekly instalments of \$15. Which way costs less and by how much?	(A) Paying by cash, \$70 (B) Instalments, \$70 (C) Paying by cash, \$130 (D) Instalments, \$130
21	Instead of dividing by 4, Paul multiplied by 4 and finished with an answer of 624. What was the correct answer?	(A) 624 (B) 2 496 (C) 39 (D) 156
22	If ¾ of Don's marbles equal 96, how many has he?	(A) 256 (B) 128 (C) 64 (D) 32
23	How many cubic centimetres are there in 1¼ litres?	(A) 1.25 (B) 12.5 (C) 125.0 (D) 1250
24	What is the perimeter of an equilateral triangle if one side is 5½cm?	(A) 16.5mm (B) 16.5m (C) 0.165m (D) 11cm
25	A sandwich shop sells 3 different types of sandwiches: ham, chicken and egg. The fractions of each which are sold are represented on the bar graph. Ham | Chicken | Egg 0 1 2 3 4 5 6 7 8 9 10 7.5 On a particular day, 120 sandwiches were sold. The number of these that were egg was	(A) 25 (B) 30 (C) 35 (D) 40
26	Which of the given possibilities has all its faces the same?	(A) cone (B) cylinder (C) square pyramid (D) cube
27	Which of the given possibilities is not a solid?	(A) sphere (B) cone (C) circle (D) cube

Questions 28-32 refer to the pie chart given which represents the percentages of a household's monthly income spent on various items. All lines passing through the centre are straight.	
28 What percentage of the income is spent on Gas?	(A) 9% (B) 16% (C) 25% (D) 20%
29 What fraction of the income is spent on "OTHERS"?	(A) ½ (B) ⅓ (C) ¼ (D) ⅕
30 What percentage of the income is spent on rent, gas and clothing altogether?	(A) 30% (B) 40% (C) 50% (D) 60%
31 It was found that the amount of the household's income spent on clothing was the same as the amount spent on food. If the household's income was reduced by a half of the present amount, what percentage of the income would be spent on clothing?	(A) 16% (B) 8% (C) 4% (D) 25%
32 If $180 was spent on electricity in the month, what amount was spent on food in the same month?	(A) $160 (B) $180 (C) $250 (D) None of these
33 Write the number one thousand less than one million.	(A) 990 900 (B) 990 000 (C) 99 900 (D) 999 000
34 A clock is 4 minutes fast when the correct time is 10 a.m. but it is losing 2 minutes every half hour. What time will it show when the correct time is 4 p.m. the same day?	(A) 4.04 p.m. (B) 4.02 p.m. (C) 3.56 p.m. (D) 3.40 p.m.
35 A tree 6 metres high casts a shadow 8 metres long. How high is a pole which casts a shadow 3 metres long?	(A) 3m (B) 6m (C) 2.5m (D) 2.25m
36 Julie mixes 4 parts of water and 1 part of cordial to make orange drink. How much drink does she make from 8 litres of cordial?	(A) 32L (B) 40L (C) 8L (D) 50L

37	On average each sample of ore from a gold mine contains 6% gold. How much gold would be expected from 2 000kg of ore?	(A) 60kg (B) 600kg (C) 12kg (D) 120kg
38	At a car factory, each assembly line produces 8 cars every 10 minutes. If there are 7 assembly lines, how many cars are produced each hour?	(A) ($\frac{10}{8}$ x 60) x7 (B) (8 x $\frac{10}{60}$) x 7 (C) (8 x $\frac{60}{10}$) x 7 (D) (8 x 10) x 7
39	A radio station broadcasts 1½ hours of news each day. If the station is on air for 24 hours a day, what fraction of its weekly listening time is devoted to news?	(A) $\frac{105}{240}$ (B) $\frac{1}{16}$ (C) $\frac{3}{4}$ (D) $\frac{3}{16}$
40	Jimmy has 4 times as much money as his sister. Between them, they have $80. How much of that was Jimmy's?	(A) $60 (B) $64 (C) $32 (D) $20
41	Through how many degrees will the big hand of a clock turn between 1 p.m. and 6 p.m. on the same day?	(A) 180 (B) 150 (C) 90 (D) None of these
42	The figure shows a regular hexagon PQRSTU inside a circle centre O, radius 6cm. It is made up of shapes 1, 2 and 3. If the perimeter of the hexagon is 36cm, what is the perimeter of shape 3?	(A) 30cm (B) 36cm (C) 24cm (D) 40cm
43	12% of a sum of money is $240. The sum of money is	(A) $2 000 (B) $2 400 (C) $200 (D) None of these

Questions 44 and 45 refer to the diagram given. The triangle ABF and the triangle CDE are equilateral triangles. BCEF is a square. The length FE is 16m.	A, B, F, x, 16m, C, E, D
44 The perimeter of the above figure is	(A) 64m (B) 96m (C) 32m (D) 128m
45 The angle x is equal to	(A) 60° (B) 90° (C) 140° (D) 150°

Selective Schools
Paper 3

1	$2^2 + 2^3$ is equal to	(A) 12 (B) 10 (C) 24 (D) 32
2	Which of the following is equal to 64.01?	(A) 64.01 + 1 (B) 64.01 - 1 (C) 64.01 x 0 (D) 64.01 ÷ 1
3	How many axes of symmetry does the star have?	(A) 1 (B) 3 (C) 4 (D) 6
4	Find the counting number which should replace the frame to make the number sentence true. 4 x □ - 288 = 52	(A) 120 (B) 115 (C) 98 (D) 85
5	Which is false?	(A) 6 x 4 ÷ 3 = 6 x (4 ÷ 3) (B) 4 ÷ 5 x 2 = 4 ÷ (5 x 2) (C) 4 x 7 x 9 = (9 x 4) x 7 (D) 27 ÷ 3 x 4 = (27 ÷ 3) x 4
6	The product 296 x 9 is the same as	(A) 2 700 - 9 (B) 2 700 - 36 (C) 2 700 - 81 (D) 2 700 - 54
7	Which of the following expressions is the same as $1\frac{3}{10}$?	(A) 3 ÷ 10 (B) 13 x 10 (C) 13 ÷ $\frac{1}{10}$ (D) 13 ÷ 10
8	My age is 61. I was married 42 years ago and graduated 7 years after that. How old was I when I graduated?	(A) 19 (B) 24 (C) 26 (D) 32
9	A schoolgirl leaves home at 7:49 a.m. and arrives back home at 3:34 p.m. How long does she spend away from home?	(A) 7 hours 45 minutes (B) 7 hours 49 minutes (C) 4 hours 15 minutes (D) 8 hours 45 minutes
10	A water tank which can hold 2 400 litres is $\frac{3}{4}$ full. If $\frac{1}{4}$ of the water is used, how many litres are left?	(A) 2 100L (B) 1 350L (C) 1 200L (D) 1 800L

11 What is the temperature difference between -15°C and 15°C?	(A) 15°C (B) 0°C (C) 30°C (D) -15°C
12 Write a simple decimal expression for $\frac{1}{10} + \frac{3}{100} + 2$.	(A) 2.31 (B) 2.13 (C) 23.1 (D) 231
13 Which is the largest of the given results?	(A) 20% of 10 (B) 2% of 100 (C) 40% of $\frac{1}{2}$ (D) 25% of 9
14 Which statement is false?	(A) All the angles of an equilateral triangle are equal in size. (B) A diagonal of a rectangle is an axis of symmetry. (C) A square has four axes of symmetry. (D) The inside (interior) angles of a triangle total 180°.
15 Mary planted $\frac{1}{2}$ of her garden with beans, and $\frac{1}{3}$ with lettuce. What fraction of her garden is not planted?	(A) $\frac{1}{5}$ (B) $\frac{4}{5}$ (C) $\frac{1}{6}$ (D) $\frac{5}{6}$
16 How fast in kilometres per hour is a car travelling if it travels 20km in 15 minutes?	(A) 100 (B) 90 (C) 80 (D) 60
17 John paid $700 for a T.V. set after he received 12.5% discount. What was the price before discount was deducted?	(A) $712.50 (B) $850 (C) $756 (D) $800
18 A girl types 3 words every 10 seconds. How many words has she typed in $3\frac{1}{2}$ minutes?	(A) 21 (B) 35 (C) 63 (D) 110
19 Find the area of the following diagram in square metres - all angles are right angles. 8m 3m 6m 2m	(A) 48 (B) 42 (C) 39 (D) 18

20	George made a mistake. Instead of multiplying 16.4m by 25, he only multiplied by 5. By how many metres was the answer wrong?	(A) 16.4 (B) 82 (C) 32.8 (D) 328
21	At a sale a microwave oven usually priced at \$210 is sold at a discount of 15%. What was the sale price?	(A) \$31.50 (B) \$178.50 (C) \$179.50 (D) \$177.50
22	Add 1½ to the difference between 2¾ and ¼.	(A) 4 (B) 3¾ (C) 3 (D) 3½
23	Milk can be bought in different quantities. Which of the given possibilities would give the best value for money?	(A) 500mL for \$1 (B) 1L for \$1.96 (C) 5L for \$9.90 (D) 3L for \$6.03
24	Which of the given fractions is equivalent to 3/4?	(A) $\frac{3-1}{4-1}$ (B) $\frac{3 \times 5}{4 \times 5}$ (C) $\frac{3+2}{4+2}$ (D) $\frac{3}{4}+\frac{4}{4}$
25	Which solid is formed from the following net? 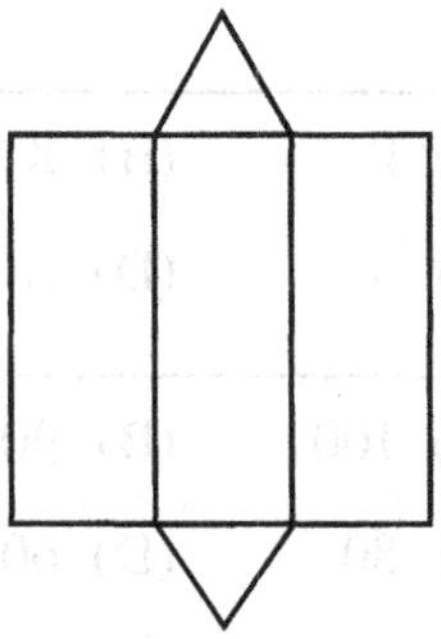	(A) cube (B) rectangular prism (C) triangular prism (D) triangular pyramid
26	Find the numbers whose sum is 25 and product is 100.	(A) 10 and 5 (B) 25 and 4 (C) 20 and 5 (D) 10 and 15
27	Mr McDonald's weekly income of \$600 is spent in the following way: 3 parts on tax, 4 parts on food, and one part on other items. How much does he pay for food from his income?	(A) \$300 (B) \$200 (C) \$100 (D) \$500

28 Complete the pattern ?	(A) (B) (C) (D)
29 2 4½ 7 A C D E B The line AB passes through the centres C, D and E of each circle. The units are in centimetres. What is the length of AB in centimetres?	(A) 13½ (B) 27 (C) 23 (D) 54
30 What is the area of this triangle in cm^2? 6cm 6cm	(A) 6 (B) 36 (C) 12 (D) 18
31 Two girls together donated \$3.50 to a charity. If one girl donated one and a half times as much as the other girl, how much was the larger donation?	(A) \$2.00 (B) \$2.10 (C) \$2.50 (D) \$2.80
32 What is the total surface area of this rectangular closed block in cm^2? 4cm 5cm 6cm	(A) 148 (B) 74 (C) 120 (D) 60
33 John, Peter and Paul went out for a night in town. They bought ice creams costing \$6 in all, had a meal costing \$10.50 each, bus fares \$2 each and movie tickets cost \$6 each. If they shared expenses equally, how much did it cost each boy?	(A) \$24.50 (B) \$23.50 (C) \$22.50 (D) \$20.50

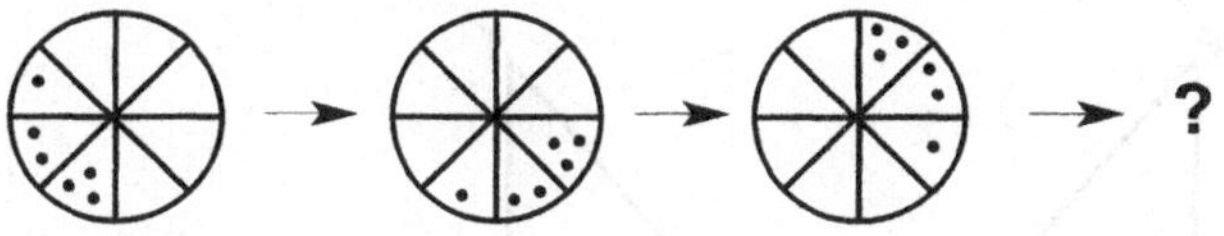

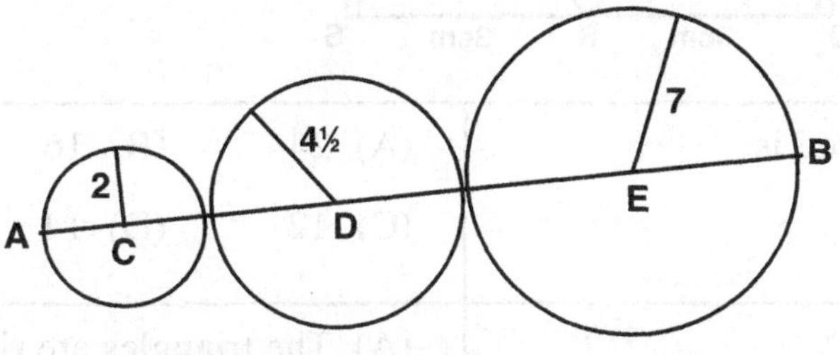

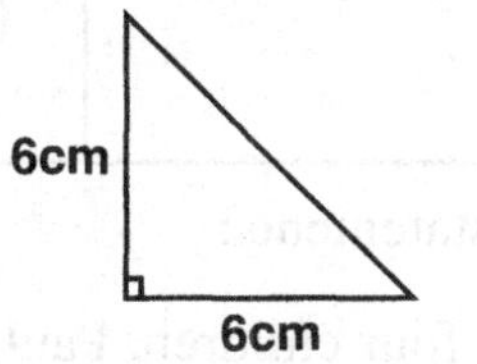

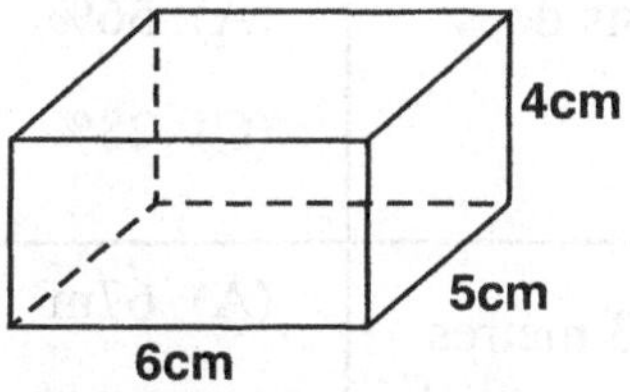

Questions 34 and 35 refer to the diagram shown. QRS is a straight line.

P, T, 4cm, 4cm, Q, 3cm, R, 3cm, S

34	The area of the above figure in cm^2 is	(A) 24 (B) 16 (C) 12 (D) 14
35	Which statement is false?	(A) The triangles are right-angled triangles. (B) The side PR is equal to the side TR. (C) Angle PRQ = angle TRS (D) Angle QPR + angle TRS = 180°

Questions 36 and 37 refer to the following statements:

A certain number of pens are shared among four children, Paul, Jane, Michelle and George. Paul has 5 more than Michelle. Jane has twice as many pens as Paul. George has the same number as Michelle.

36	If Jane has 14 pens, what is the total number of pens?	(A) 111 (B) 45 (C) 35 (D) 25
37	What percentage of the total number of pens does Paul have?	(A) 56% (B) 4% (C) 28% (D) 16%
38	A room 5 metres long, 4 metres wide and 3 metres high is to be painted. There are two windows which are 2.5 square metres each in area and there is one door which is 2 square metres in area. What is the area of the walls and the ceiling to be painted?	(A) $67m^2$ (B) $69.5m^2$ (C) $87m^2$ (D) $74m^2$
39	It costs 36c to run a shower for an hour. How much does it cost in cents for a 4 minute shower?	(A) 6⅔ (B) 2.4 (C) 540 (D) 144
40	Add the largest and smallest numbers in the set 0.81, 0.80, 0.79, 0.8, 0.78	(A) 1.59 (B) 1.61 (C) 1.58 (D) 1.60

41	A rectangular beam of timber is 16 metres long, 80cm wide and 50cm deep. If a piece containing 2 cubic metres is cut from the beam perpendicular to its length, what length would be left?	(A) 9m (C) 11m	(B) 10m (D) 12m
42	11 - 0.01 is equal to	(A) 11.19 (C) 10.91	(B) 10.99 (D) 10.9
43	Find 75% of 75.	(A) 56¼ (C) 48	(B) 50 (D) 25
44	The side of a regular hexagon is 5cm long. A square has the same perimeter as the hexagon. The side of the square is	(A) 7½cm (C) 56¼cm	(B) 6¼cm (D) 30cm
45	How many squares of side 0.2cm can be cut from a square of side 4cm?	(A) 40 (C) 400	(B) 20 (D) 80

Selective Schools
Paper 4

1	The answer to 0.6 x 7 is	(A) 0.042 (B) 0.42 (C) 4.2 (D) 42
2	Which of the given results has the smallest value?	(A) ⅖ (B) 0.26 (C) 2.06 (D) ¼
3	14 x 2 + 2 x 15 is	(A) 150 (B) 60 (C) 58 (D) 48
4	In the following operations, the missing signs are 4 892 ☐ 3 127 = 1 765 1 765 ☐ 3 127 = 4 892	(A) - and - (B) + and + (C) x and x (D) - and +
5	One half of 80 is the same as one fifth of	(A) 100 (B) 200 (C) 300 (D) 400
6	If the zero is left out, which of the given numbers will remain unchanged?	(A) 6.209 (B) 20.7 (C) 7.890 (D) 790
7	A girl went swimming at 11:35 a.m. and swam for 3 hours 45 minutes. At what time did she stop swimming?	(A) 2:15 p.m. (B) 2:45 p.m. (C) 3:15 p.m. (D) 3:20 p.m.
8	Coloured marbles are placed in a row in the order red, yellow, pink, blue. What colour would the 58th marble be?	(A) red (B) yellow (C) pink (D) blue
9	What digit should replace the ☐ so that the number 12 ☐ 7 is divisible by 9?	(A) 7 (B) 5 (C) 3 (D) 8
10	What number should replace the ☐ in order to make the number sentence below true? 16 x ☐ = 9 x 5 + ☐	(A) 5 (B) 3 (C) 6 (D) 16
11	In 5 years time I will be 3 times as old as my son. In 14 years time, the sum of our ages will be 62. Today the sum of our ages is	(A) 34 (B) 36 (C) 40 (D) 56
12	A car travels 6km in 5 minutes. At this speed, how far in kilometres would it travel in one hour?	(A) 30 (B) 60 (C) 72 (D) 80

13	A triangle has sides of length 7cm, 14cm and 15cm. An equilateral triangle has the same perimeter. What is the length in centimetres of each side of the equilateral triangle?	(A) 7 (B) 14 (C) 12 (D) 15
14	The digits of the number 2594 are arranged in descending order and then in ascending order. What is the difference between the resulting numbers?	(A) 8 083 (B) 7 083 (C) 6 083 (D) 5 803
15	What is the quotient when 0.1 is divided by 2?	(A) 5 (B) 0.5 (C) 0.05 (D) 0.005
16	$(11 - 11) + \frac{11 \times 11}{11 \div 11} - (11 + 11) =$	(A) 121 (B) 242 (C) 22 (D) 99
17	Six more than twice the sum of a certain number and 5 equals 30. The certain number must be	(A) 7 (B) 10½ (C) 9 (D) 11
18	On a map the scale is 1cm = 25km. How long is a road that is shown on the map as 7¼ cm?	(A) 180.25km (B) 182.05km (C) 182.5km (D) 181.25km
19	A plane flying south turned anti-clockwise through 135°. In what direction is it now flying?	(A) NW (B) SW (C) NE (D) SE
20	How many hectares are there in a block of land 0.5 km long and 0.2 km wide?	(A) 1ha (B) 0.01ha (C) 10ha (D) 100ha
21	Trees alongside a highway are spaced at a distance of 20 metres apart. What is the maximum number of trees that will lie in 1km?	(A) 20 (B) 50 (C) 51 (D) 100
22	How many cm^3 of oil are required to fill a rectangular tank 7m deep, 8m long and 10m wide?	(A) 560 (B) 56 000 (C) 5 600 000 (D) 560 000 000
23	The product of three numbers is 2 688. Two of them are 14 and 16. What is the third number?	(A) 5 (B) 10 (C) 15 (D) 12
24	In a school of 650 boys, 130 of them were unable to pass a fitness test. What percentage passed?	(A) 20% (B) 40% (C) 60% (D) 80%

25	How many tiles each measuring 20cm x 25cm are needed to cover a surface measuring 7m x 7m?	(A) 980 (B) 98 (C) 9.8 (D) 0.98
26	Which is the larger: 3% of 100 or 100% of 3?	(A) 3% of 100 (B) 100% of 3 (C) they are equal (D) None of these
27	The circumference of John's bicycle wheel is 1.61 metres. How many metres will the bicycle travel in 25 revolutions?	(A) 20.25 (B) 25.25 (C) 35.25 (D) 40.25
28	How many times is 100 contained in 10 000 000?	(A) 5 (B) 10 000 (C) 100 000 (D) 1 000 000
29	At 7:00 a.m. the temperature was 9.6°C and at 1:00 p.m. the temperature was 24.9°C. What was the average hourly increase?	(A) 3.5°C (B) 2.55°C (C) 2.25°C (D) 2.15°C
30	Which of the given decimals is closest to $\frac{7}{10}$?	(A) 0.685 (B) 0.71 (C) 0.07 (D) 7.0
31	An empty swimming pool has a capacity of 48 000 litres. How long will it take before the pool is ¾ full if it is being filled at a rate of 900 litres per minute?	(A) 2 hours (B) 1 hour (C) 40 minutes (D) 20 minutes
	Questions 32 and 33 refer to the given figure.	
32	The letter E as shown in the diagram, is formed using squares of side 1 metre. What is the perimeter of the figure in metres?	(A) 31 (B) 32 (C) 33 (D) 34
33	What is the area of this figure in square metres?	(A) 12 (B) 16 (C) 18 (D) 20

34	Peter went shopping. How much did it cost him if he bought the following items? 2kg of apples at $1.20 a kilogram 3 packets of meat at $2.45 a packet 16 bread rolls at 15 cents each 5 litres of milk at 45 cents per litre	(A) $14.40 (B) $28.80 (C) $56.25 (D) None of these
35	A boy does 7 subjects. His average mark for 6 subjects was 72 and his mark in the next subject was 100. What is his average now?	(A) 74 (B) 75 (C) 76 (D) 77
36	Mary's step is 40cm and Jim's step is 50cm. How many more steps does Mary take than Jim in walking 2 kilometres?	(A) 1 000 (B) 100 (C) 10 (D) 10 000
37	An author receives 20% in royalties on the sale of his best sellers. How much did he receive if he sold 160 000 books at $13.80 each?	(A) $4 416 (B) $44 160 (C) $441 600 (D) $73 680
38	If ⅔ of a telegraph pole is below the ground and 15 metres is above the ground, what is the total length of the pole?	(A) 45m (B) 20m (C) 15m (D) 30m
39	A girl earns $1.60 per hour as a babysitter. How many complete hours would she have to work in order to earn $20?	(A) 11 (B) 12 (C) 13 (D) 14
40	If 40% of a number is 64, then 70% of the number is	(A) 112 (B) 94 (C) 36.6 (D) 448

Questions 41 to 43 refer to the given graph:

Cost of Aluminimum and Lead

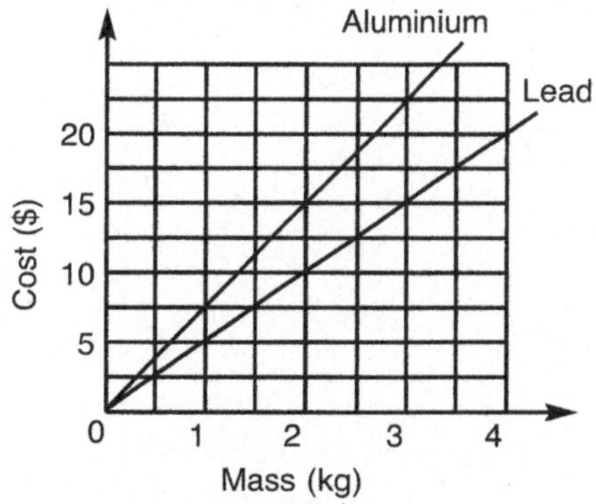

41	The cost of 2kg of lead and 2kg of aluminium is:	(A) $15 (B) $20 (C) $25 (D) $30
42	How many kilograms of aluminium can be bought with $15 000?	(A) 200 (B) 2 000 (C) 20 000 (D) 200 000

43	A company purchased $100 000 worth of aluminium and lead. If the mass of the aluminium was 10 000kg, how many kilograms of lead did the company purchase?	(A) 10 000kg (B) 11 000kg (C) 5 000kg (D) 13 000kg
44	A large container is 8 metres long, 3 metres high and 5 metres wide. How many boxes 2 metres long by 50cm wide and 20cm high could be packed into the large container?	(A) 1 200 (B) 120 (C) 600 (D) 300
45	What is the most likely number to complete the pattern 1, 11, 3, 9, 5, 7, 7, □?	(A) 6 (B) 8 (C) 5 (D) 9

Selective Schools
Paper 5

1	7 - 3.48 is equal to	(A) 10.48 (B) 3.51 (C) 4.62 (D) 3.52
2	6 x 4.37 is equal to	(A) 0.262 (B) 27.22 (C) 2.62 (D) 26.22
Questions 3-4: A rectangle is 250cm long and 50cm wide.		
3	The perimeter in metres is	(A) 600 (B) 300 (C) 60 (D) 6
4	The area in square metres is	(A) 12 500 (B) 1 250 (C) 1.25 (D) 12.5
5	The difference between one million dollars and one million cents is	(A) \$999 990 (B) \$999 900 (C) \$999 000 (D) \$990 000
6	The square root of the sum of the squares of 6 and 8 is	(A) 14 (B) 10 (C) 100 (D) $\sqrt{28}$
7	A closed cube has a volume of 27cm³. The total surface area in cm² is	(A) 9 (B) 54 (C) 81 (D) 486
8	0.32 ÷ 4 is equal to	(A) 0.08 (B) 0.8 (C) 8 (D) none of these
Questions 9-10: The length of the circumference of a circle can be taken as three and one seventh times the length of the diameter.		
9	A wheel of a cycle has a diameter of 70cm. The length, in cm, of the circumference of the wheel is	(A) 220 (B) 2.2 (C) 22 (D) 4.4
10	The number of revolutions of this wheel in cycling a distance of 110m is:	(A) 500 (B) 5 000 (C) 50 (D) 5
11	The average of the numbers 2, □, 7, 6 is 4.5. The value of □ is	(A) 10.5 (B) 19.5 (C) 4 (D) 3

12	A recipe says that 10kg of fruit need 8kg of sugar. The number of kilograms of fruit than can be made with 6kg of sugar is	(A) $13\frac{1}{3}$ (B) $4\frac{4}{5}$ (C) $7\frac{1}{2}$ (D) $6\frac{1}{2}$
13	As a percentage, $\frac{3}{100} + \frac{6}{10}$ is	(A) 36 (B) 63 (C) 3.6 (D) 0.63
	Questions 14-15: An empty room is 7m long, 5m wide and 3m high.	
14	Ignoring doors and windows, the area of the four walls in m^2 is	(A) 105 (B) 72 (C) 36 (D) 142
15	If $1m^3$ = 1000 litres, then the capacity of the room in litres is	(A) 105 (B) 15 000 (C) 105 000 (D) 72 000
16	If $2\frac{1}{2}$ litres of juice cost \$1.80, then $1\frac{1}{2}$ litres cost	(A) 96c (B) \$1.08 (C) \$1.26 (D) \$1.20
17	Which of the given values is the best approximation to $\frac{75.32 \times 138.7}{12.62 \times 81.46}$?	(A) 0.1 (B) 1 (C) 10 (D) 0.001
18	4 586 persons are each given a $\frac{1}{2}$mL dose of vaccine. The number of litres of vaccine used was	(A) 4.586 (B) 22.93 (C) 2.293 (D) 45.86
19	The number represented by four thousand seven hundred and nine units is	(A) 479 (B) 4 079 (C) 4 709 (D) 40 709
20	A set of 5 measurements has an average of 13, whilst another set of 7 measurements has an average of 10. The average of the 12 measurements is	(A) 23 (B) 11.5 (C) 11.25 (D) 4.02
21	How many multiples of 3 are there less than 40 which are also divisible by 4?	(A) 13 (B) 6 (C) 10 (D) 3
22	The operation P between two numbers means to find four times the sum of these numbers. The value of 5P6 is	(A) 11 (B) 48 (C) 44 (D) 15

23	A pie chart is drawn to represent 3 items. The angles of two of the sectors are 107° and 208°. The third sector as a percentage of the whole pie chart is	(A) 25 (B) 87.5 (C) 12.5 (D) 62.5
24	It takes 50 minutes to drive to town. When should I leave if I have to be in town at 9.35 a.m.?	(A) 8:45 a.m. (B) 9:05 a.m. (C) 8:55 a.m. (D) 8:25 a.m.
25	28 + 0.1 x 40 =	(A) 68 (B) 32 (C) 1 124 (D) 28.4
26	The number halfway between 0.4 and 1.68 is	(A) 0.68 (B) 1.04 (C) 1.28 (D) 1.08
27	A quadrilateral with exactly four lines of symmetry must be a	(A) rectangle (B) parallelogram (C) square (D) rhombus
28	A scout walks in a direction of NE. In what direction must he walk in order to return to his starting point?	(A) SE (B) NW (C) SW (D) NE
29	Write 300 as a fraction of 60 000.	(A) $\frac{1}{20}$ (B) $\frac{1}{40}$ (C) $\frac{1}{100}$ (D) $\frac{1}{200}$
30	An olive tree was planted 4 years and 5 months before the olives were picked in March 1991. The tree was planted in:	(A) November 1986 (B) November 1987 (C) October 1986 (D) October 1987
31	If the length of a rectangle is doubled and the breadth tripled, then the area of the rectangle has been multiplied by:	(A) 2 (B) 3 (C) $2\frac{1}{2}$ (D) 6
32	A pill bottle containing 16 pills has a total mass of 140g. When the bottle contains 10 pills its total mass is 125g. The mass of the bottle in grams is:	(A) 92 (B) 100 (C) 15 (D) 110
33	20% of $8 is:	(A) $16 (B) 16c (C) $160 (D) $1.60
34	The group in which the numbers are arranged in increasing size is:	(A) $\frac{3}{4}$, 0.76, 74% (B) 74%, $\frac{3}{4}$, 0.76 (C) 0.76, $\frac{3}{4}$, 74% (D) 0.76, 74%, $\frac{3}{4}$

35	The sum of all the even divisors of 32 (including 32 itself) is:	(A) 60 (B) 63 (C) 54 (D) 62
36	The value of 100 - 99 +98 -97 + ... - 3 + 2 - 1 is:	(A) 50 (B) 49 (C) 51 (D) 99
37	The square shown is a magic square. This means that the sum of the numbers in any row, column or diagonal must be the same. The value of X is: | | | 12 | | 9 | | 13 | | | X | 8 |	(A) 14 (B) 7 (C) 11 (D) 15
38	Neil, Jane and Mary each have $30. How much money should Jane give to Neil and Mary so that Neil has $5 more than Jane and Mary has $1 less than Neil?	(A) $5 to Neil, $1 to Mary (B) $5 to Neil, $4 to Mary (C) $3.50 to Neil, $2.50 to Mary (D) $2 to Neil, $1 to Mary
39	Write the number which is one hundred less than one million.	(A) 990 000 (B) 99 900 (C) 999 900 (D) 990 900
40	20 marks out of 25 is what as a percentage?	(A) 25% (B) 40% (C) 80% (D) 60%
41	Complete this pattern → → → ?	(A) (B) (C) (D)
42	Which two numbers have their sum as 11 and product as 24?	(A) 4 and 10 (B) 4 and 6 (C) 2 and 9 (D) 3 and 8
43	Which one of the given possibilities does not equal 0.4?	(A) $\frac{2}{5}$ (B) 40% (C) $\frac{4}{10}$ (D) 4%

44	A ten dollar note is approximately 15.5cm long. About how far will 1 000 ten dollar notes reach if laid end to end?	(A) 15.5m (B) 15.5km (C) 1.55km (D) 155m
45	An urn is half full of water. If 20 litres of water is added to it, the urn is then three quarters full. How many litres does the urn hold when full?	(A) 5 (B) 40 (C) 100 (D) 80

Selective Schools
Paper 6

1	4% of 1 500 equals	(A) 60 (B) 600 (C) 6 (D) 375
2	Which of the given possibilities is the closest approximation to 3.97 x 2.134?	(A) 80 (B) 8 (C) 0.08 (D) 0.8
3	The distance between two buildings is 350m. This distance is represented on a map by a length of 7mm. The scale used on the map is	(A) 1 : 50 (B) 1 : 500 (C) 1 : 5 000 (D) 1 : 50 000
4	A car is travelling at a speed of 78km/h. The distance it covers in 20 minutes is	(A) 1 560km (B) 19.5km (C) 13.6km (D) 26km
5	In the following diagram the shaded area in m^2 is 12m 6m 8.5m 4.5m	(A) 129 (B) 20 (C) 10 (D) 75
6	The value of 0.03 x 3 is	(A) 0.06 (B) 0.09 (C) 0.9 (D) 0.6
7	Four identical discs are marked with the digits 1,3, 5, 7 respectively. The four discs are placed in a row so that the digits form a number. How many different numbers are possible, each number starting with 5?	(A) 24 (B) 12 (C) 6 (D) 16
8	The value of 0.2 ÷ 5 is	(A) 0.04 (B) 0.4 (C) 4 (D) 40
9	93cm of rain fell in the period January to November inclusive (i.e. including January and November). The average monthly rainfall for the whole year was 9cm. How many cm of rain fell in December?	(A) 84 (B) 201 (C) 6 (D) 15
10	As a percentage $\frac{17}{25}$ is	(A) 68 (B) 34 (C) 51 (D) 85

Questions 11-12: The lengths of the sides of a triangle are 4.53cm, 5.71cm and 6.2cm.	
11 The perimeter of the triangle in centimetres is	(A) 16.26 (B) 15.44 (C) 16.44 (D) 15.144
12 The difference in length between the shortest and the longest sides, in millimetres, is	(A) 1.67 (B) 16.7 (C) 0.167 (D) 17.7
13 It costs $135 to hire a bus. In addition, there is a charge of 90c per kilometre travelled. On a journey of 50km, there are 40 passengers. If the total cost of the trip is shared equally by the passengers, then the cost per passenger is	(A) $12.80 (B) $14.63 (C) $1.25 (D) $4.50
14 Brass is made up of copper and zinc so that for each 13 parts of copper there are 7 parts of zinc. How many kilograms of zinc would there be in 1 tonne of brass?	(A) 650 (B) 35 (C) 350 (D) 3 580
15 In the number 632, the 6 has a value that is	(A) three times the value of the 2 (B) twice the value of the 3 (C) thirty times the value of the 2 (D) twenty times the value of the 3
16 The decimal numeral represented by $4 + \frac{3}{10} + \frac{7}{100}$ is	(A) 4.37 (B) 4.73 (C) 43.7 (D) 437
17 The number of grams in 2.3kg is	(A) 23 (B) 2 300 (C) 230 (D) 23 000
18 If 90% of a certain number is 360, the number is	(A) 400 (B) 40 (C) 4 000 (D) 324
19 Sheets of paper cost $3\frac{1}{2}$ cents each. The number of sheets which can be bought for $1.54 is	(A) 440 (B) 539 (C) 48 (D) 44
20 An article costing 24 cents is sold for 30 cents. The gain as a fraction of the selling price is	(A) $\frac{1}{5}$ (B) $\frac{1}{4}$ (C) $\frac{4}{5}$ (D) $\frac{5}{4}$

21	Jose won $1 000 000 in a lottery. He decided that he would spend $1 a second for the next 10 years. But he was disappointed. All his money was gone after approximately	(A) 1 day (B) 10 days (C) 10 weeks (D) 1 year
22	A rectangular tank 3m x 4m x 5m is full of water. If 1 litre of water weighs 1kg, what is the weight of the water in tonnes?	(A) 600 (B) 16 (C) 6 (D) 60
23	The area in ha of a rectangular farm 2km long and 1 600m wide is	(A) 320 (B) 3 200 (C) 32 (D) 32 000
24	⅛ of 1 litre in millilitres is	(A) 12.5 (B) 125 (C) 1 250 (D) 1.25
25	In a class of 48 students, 36 are girls. The fraction of boys in the class is	(A) ¾ (B) ¼ (C) ⅓ (D) ⅔
26	$36 is shared between A, B and C so that A received 4 parts, B received 3 parts and C received 2 parts. What did C receive?	(A) $12 (B) $10 (C) $8 (D) $6
27	A car uses 500mL of petrol to travel 5.5km. How many litres of petrol would the car use to travel 218km? Approximately	(A) 19litres (B) 20litres (C) 21litres (D) none of these
28	Which of the following quadrilaterals has exactl two axes of symmetry.	(A) trapezium (B) parallelogram (C) square (D) rhombus
29	Orange juice is sold in 3 litre casks. If paper cups hold 160mL, how many full cups of orange juice can you get from a cask?	(A) 190 (B) 19 (C) 20 (D) 18
30	There are six more squares than triangles on a page. If there are 42 such diagrams altogether, how many squares are there?	(A) 18 (B) 24 (C) 36 (D) 30

31	In the sketch, half the smaller square overlaps the larger square, whose side is 9cm. If the area of the figure ABCDEFGH is 89cm², then the side of the smaller square in centimetres is A B G H C D F E	(A) 4 (B) 3 (C) 2 (D) 5
32	If you were to write down all the counting numbers from 1 to 120, how many times would the digit '7' occur?	(A) 12 (B) 13 (C) 21 (D) 22
33	The angle between the hands of a clock at half past two is	(A) 105° (B) 120° (C) 90° (D) 100°
34	A number is divided by each of the numbers 6, 8, 9, 24. In each case it leaves a remainder of 1. The smallest such number would be	(A) 10 369 (B) 49 (C) 145 (D) 73
35	A rectangle measures 16cm x 9cm. The side of a square of equal area to the rectangle is	(A) 10 cm (B) 12 cm (C) 14 cm (D) 12½cm
36	3 + 3.3 equals:	(A) 3.6 (B) 6.3 (C) 33.3 (D) 3.33
37	$\frac{3 \text{ x } 4 \text{ x } 20}{2 \text{ x } 5 \text{ x } 8 \text{ x } 12}$ is equal to:	(A) 2 (B) ½ (C) 4 (D) ¼
38	Today, after assembly, classes did not start until 9:12 a.m. How many hours and minutes were there until the final siren at 3:10 p.m.?	(A) 5 hours 58 minutes (B) 6 hours 58 minutes (C) 6 hours 2 minutes (D) 12 hours 22 minutes
39	Write the number which is one hundred thousand one hundred less than two million.	(A) 1 899 900 (B) 1 899 090 (C) 1 890 000 (D) 1 899 000
40	Which is the smallest of the given possibilities?	(A) 3.3 (B) 3.03 (C) 3.30 (D) 3.33

41	The total fare of 2 adults and 3 children is $16. If the children's fare is $1.80 each, how much is each adult fare?	(A) $10.60 (B) $5.40 (C) $6.20 (D) $5.30
42	Arthur pays for an article with a $2 coin and receives three different silver coins as change. What is the greatest amount of money the article could have cost?	(A) 35c (B) $1.92 (C) $1.85 (D) $1.65
43	Which of these numbers is prime?	(A) 57 (B) 49 (C) 51 (D) 53
44	1 - $^{7}/_{1000}$ is:	(A) 0.3 (B) 0.97 (C) 0.93 (D) 0.993
45	2 U 3 3 4 x 1 0 V 2 7 W 9 8 9 4 2 In the multiplication shown, the digits U, V, W are respectively:	(A) 8, 6, 5 (B) 6, 5, 9 (C) 6, 5, 8 (D) none of these

Selective Schools
Paper 7

1	45cm as a percentage of 1m is	(A) 0.045% (B) 0.45% (C) 4.5% (D) 45%
2	$165 is divided among 3 girls so that for each $2 obtained by the first girl, the second girl gets $3 and the third girl gets $6. The first girl's share of the $165 is	(A) $30 (B) $45 (C) $90 (D) $15
3	430.2 ÷ 9 is	(A) 47.8 (B) 4.78 (C) 478 (D) 0.478
	Questions 4-5: On a map, 1mm represents a distance of 20km.	
4	The distance in km represented by a length of 2.25cm is	(A) 4.5 (B) 45 (C) 450 (D) 4 500
5	What length in centimetres on the map would represent a distance of 700km?	(A) 35 (B) 3.5 (C) 350 (D) 0.35
6	0.7 of $23 is	(A) $1.61 (B) $16.10 (C) $161 (D) $30
	Questions 7-8 refer to the diagram shown. The given figure is symmetrical and all angles are right angles.	
7	The perimeter of the figure in centimetres is	(A) 48 (B) 40 (C) 36 (D) 24
8	The area of the given figure in cm² is	(A) 96 (B) 72 (C) 78 (D) none of these

9	**O** is the centre of the circle. In the diagram, the interval AB is called a	(A) radius (B) diameter (C) chord (D) arc
10	In the figure, O is the centre of the circle. the shaded area is a	(A) segment (B) semicircle (C) sector (D) quadrant
11	An isosceles triangle is a triangle with	(A) no sides equal (B) 2 sides equal (C) 3 sides equal (D) no angles equal
12	The cost for a motor trip of 73km was \$14.60. What should be the cost at the same rate for a journey of 65km?	(A) \$13.00 (B) \$130 (C) \$116.80 (D) \$14.15
	Questions 13-14: An empty room is 15m long, 12m wide and 10m high	
13	The area of the 4 walls and ceiling in metres^2 is	(A) 720 (B) 900 (C) 540 (D) 1 800
14	The volume in metres^3 of air it contains is	(A) 1 800 (B) 37 (C) 330 (D) 270
15	In figures, one million, twenty four thousand and thirty is	(A) 1 024 003 (B) 1 024 030 (C) 1 240 030 (D) 124 300
16	A satellite orbits the earth once every 90 minutes. How many complete revolutions does it make in 24 hours?	(A) 16 (B) 36 (C) 960 (D) 3
17	The next term in the sequence $\frac{1}{2}$, 1, $\frac{9}{8}$, 1, $\frac{25}{32}$, $\frac{9}{16}$, □ is probably	(A) 1 (B) $\frac{9}{16}$ (C) $\frac{49}{128}$ (D) none of these

18	If 0.0003 x □ = 30, then □ is	(A) 10 (B) 100 (C) 1 000 (D) 100 000
19	For what value of □ is 57 x 86 = 57 x 80 + 57 x □	(A) 166 (B) 60 (C) 16 (D) 6
20	How many pieces of string 34cm long can be cut from a piece of string which is 3 metres long?	(A) 10 (B) 9 (C) 8 (D) 7

Questions 21-23 refer to the following table.

The marks in a test out of 5 are shown in the table below:

MARKS	Number of students who received those marks
0	2
1	3
2	6
3	5
4	3
5	1

21	The number of students who got 80% in the test was	(A) 6 (B) 5 (C) 3 (D) 2
22	The number of students who got more than 2 marks is	(A) 6 (B) 5 (C) 9 (D) 20
23	What percentage of students got more than 1 and less than 5?	(A) 90 (B) 80 (C) 60 (D) 70
24	The length of a rectangle is 3m less than its breadth. If its area is 40m^2 then its perimeter must be	(A) 43m (B) 64m (C) 26m (D) 25m
25	What is 8 x 8 x 8 rounded off to the nearest hundred?	(A) 100 (B) 510 (C) 400 (D) 500
26	Our grandfather died on his 80th birthday. About how many days did he live?	(A) 280 000 (B) 28 000 (C) 290 000 (D) 29 000

27	A multi-storied office building has rectangular floors 25m long and 16m wide. How many floors does it have if there are 5 600m^2 of floor space?	(A) 10 (B) 14 (C) 9 (D) 7
28	There is enough food in a pig pen to feed 14 pigs for 16 days. For how many days should this amount of food feed 8 pigs?	(A) 28 (B) 16 (C) 26 (D) $^{64}/_{7}$
29	There are 12 girls and 8 boys in our class. If the girls averaged 64 in a test whilst the whole class averaged 60 in this test, what did the boys average?	(A) 62 (B) 48 (C) 56 (D) 54
30	The diagram shows the net of a cube of side 10cm. What is the sum of the lengths of the edges of the cube?	(A) 140cm (B) 190cm (C) 120cm (D) 600cm
31	In the sketch, not drawn to scale, the rectangle ABCD is divided into four smaller rectangles whose areas in cm^2 are shown. A B 36 63 20 35 D C If the lengths in centimetres of the sides of the rectangles are whole numbers, what is the perimeter of the rectangle ABCD?	(A) 50cm (B) 25cm (C) 28cm (D) 72cm
32	Each pair of adjoining spokes of a wheel meet at an angle of 24°. How many spokes are there?	(A) 12 (B) 15 (C) 7.5 (D) 30
33	A 1m square sheet of cardboard is cut into the maximum number of squares of side 1mm. If these squares could be laid side by side, how far would they stretch in metres?	(A) 100 (B) 1 000 000 (C) 1 000 (D) 10 000
34	School goes from 9 a.m. to 3 p.m. There is a morning break of 20 minutes and lunch takes 1 hour. The total time for classes each day is	(A) 4h 40min (B) 5h 20min (C) 4h 20min (D) 4h 50min
35	Find the sum of all the two digit numbers greater than 10 such that the tens digit is one less than the units digit.	(A) 476 (B) 414 (C) 486 (D) 404

36	The value of (10 -7) x (10 + 7) + (16 - 4) x (16 + 4) is	(A) 291 (B) 1 740 (C) 1 260 (D) 788
37	In the sketch below, the little squares are all the same size and the area of the whole rectangle is 1 square unit. The area of the shaded part in square units is	(A) $\frac{2}{15}$ (B) $\frac{1}{3}$ (C) $\frac{2}{5}$ (D) $\frac{3}{8}$
38	Which of the given expressions is equal to 3 x (5 + 7)?	(A) 5 x 3 + 5 x 7 (B) 7 x 3 + 7 x 5 (C) 3 x 5 + 3 x 7 (D) 15 + 7
39	The number 313 is exactly divisible by	(A) 3 (B) 5 (C) 11 (D) none of these
40	When the number 168 is rotated half a turn, it becomes 891. Which of the following numbers when rotated half a turn is decreased by 12?	(A) 81 (B) 69 (C) 16 (D) 98
41	A rectangular slab measuring 6m x 4m is 100mm thick. What is the volume of the slab in m^3?	(A) 24 (B) 2.4 (C) 0.24 (D) 240
42	4.86 x 0.79 is approximately equal to	(A) 4 (B) 0.4 (C) 0.04 (D) 400
43	32% expressed as a fraction is	(A) $\frac{1}{3}$ (B) $\frac{6}{25}$ (C) $\frac{8}{25}$ (D) $\frac{5}{32}$
44	What percentage of whole numbers from 2 to 21, both included, are exact multiples of 3?	(A) $\frac{19}{3}$% (B) 35% (C) 42% (D) 50%
45	A rectangle has a perimeter of 32cm and an area of $28cm^2$. What are its dimensions in centimetres?	(A) 4 and 7 (B) 1 and 28 (C) 2 and 14 (D) 56 and $\frac{1}{2}$

Selective Schools
Paper 8

1	In the number 474, the difference in the value of the 4's is	(A) 0 (B) 99 (C) 396 (D) 360
2	Of the given results, the best approximation to $58.9 \times \frac{63.4}{6.47} \times \frac{0.41}{0.38}$ is	(A) 0.6 (B) 6 (C) 60 (D) 600
3	A student obtains a mark of 29 out of 40. As a percentage, this is	(A) 62½% (B) 75% (C) $\frac{29}{40}$% (D) 72½%
4	A cooking book says that to roast a joint, the cook should allow 40 minutes for each kilogram, and then allow an additional half-hour. Following these instructions, the time required to roast a joint weighing 2½kg is	(A) 2h (B) 2h 10min. (C) 2h 20min. (D) 1h 30min.

Questions 5 - 8: This table shows how to use the operation Ø on the numbers 1, 2, 3, 4, 5.
Use this table to answer the following questions.

Ø	1	2	3	4	5
1	1	2	3	4	5
2	2	4	0	2	4
3	3	0	3	0	3
4	4	2	0	4	2
5	5	4	3	2	1

5	5 Ø 2 is equal to	(A) 1 (B) 2 (C) 3 (D) 4
6	(5 Ø 2) Ø 3 is equal to	(A) 0 (B) 1 (C) 2 (D) 3
7	The value of □ so that 2 = □ Ø 3 is	(A) 2 (B) 3 (C) 4 (D) none of these
8	The value of the number □ so that □ Ø □ = 1 is	(A) 1 only (B) 1 and 2 (C) 1, 2 and 3 (D) 1 and 5
9	If you add the number of axes of symmetry of a regular hexagon to the number of axes of symmetry of an isosceles triangle, the result is	(A) 7 (B) 5 (C) 9 (D) 4

10	A rectangular paddock is of area 7½ha. One side is 600 m long. Then the other side in metres is	(A) 125 (B) 12½ (C) 4 500 (D) $\frac{1}{80}$
11	An obtuse angle is an angle which is	(A) greater than 90°. (B) greater than 90° but less than 360°. (C) less than 90°. (D) greater than 90° but less than 180°.
12	Which is the best description of a square? A square is a quadrilateral with	(A) 4 equal sides. (B) 4 right angles. (C) the opposite sides parallel. (D) 4 equal sides and 4 right angles.
13	A pile of 200 sheets of paper is 4.8cm thick. In mm, the thickness of each sheet is	(A) 24 (B) 2.4 (C) 0.24 (D) 0.024
14	A rectangular prism is 9cm long, 8cm wide and 3cm high. A cube has the same volume as the prism. The side of the cube in centimetres is	(A) 4 (B) 5 (C) 6 (D) 8
15	3 x 0.02 x 4 is equal to	(A) 0.24 (B) 0.024 (C) 0.0024 (D) 0.009
16	357 619 rounded off to the nearest thousand, is	(A) 357 000 (B) 358 000 (C) 357 600 (D) 357 700
17	Which expression is not equal to 36?	(A) (10 - 1) x 4 (B) 80 ÷ 2 -4 (C) 3 x 8 + 2 x 6 (D) 8 + 4 x 3
18	How many square centimetres are in 1 square metre?	(A) 100 (B) 1 000 (C) 10 000 (D) 100 000
19	A truck travels 7km for each litre of fuel. If fuel costs 89c per litre, how much in fuel would a journey of 203km cost? Approximately	(A) $24 (B) $26 (C) $28 (D) $30

20	If □ = 3 and △ = 7 find the value of □ + △ x (△ - □)	(A) 40 (B) 49 (C) 67 (D) 31
21	When a wheel revolves once it is said to turn through 360°. A wheel revolves (turns) through 1620°. How many complete revolutions does it make?	(A) 4.5 (B) 5 (C) 4 (D) 6
22	I bought 9 apples at 23c each and 7 oranges at 19c each. How much change would I have from $10?	(A) $7.60 (B) $7.40 (C) $6.60 (D) $6.40
23	A rectangular prism measures 6cm by 10cm by 8cm. The prism is cut into cubes of side 2cm. The number of such cubes will be	(A) 60 (B) 48 (C) 80 (D) 64
24	A car park advertises $10 for first hour - or part thereof $5 for each subsequent hour - or part thereof The cost of a 4½ hour stay in the car park is	(A) $25 (B) $30 (C) $35 (D) $40
25	Which one of the given expressions is another way of writing 2000 + 800 + 50 + 7?	(A) 2 857 (B) 28 507 (C) 28 057 (D) 20 857
26	Which number(s) on the number line are 3 times as far from 13 as they are from 5?	(A) 1 only (B) 11 only (C) 1 and 7 (D) 11 and 17
27	Through what angle would the minute hand of a watch move from 9 a.m. to 10:15 a.m.?	(A) 75° (B) 115° (C) 340° (D) 450°
28	A clock is set correctly at 2 p.m. It loses 3 minutes every hour. What will the clock read when the correct time is 11 a.m. the next day?	(A) 10:03 a.m. (B) 10:00 a.m. (C) 12:03 p.m. (D) 9:57 a.m.
29	A certain number of consecutive odd numbers, starting at 1 are added together. Their sum is 576. How many odd numbers were added?	(A) 576 (B) 24 (C) 36 (D) 40

30 The table shows the possible sums of the number of spots uppermost when two dice are tossed.
What is the most likely sum that can occur?

(A) 6 (B) 8
(C) 7 (D) 12

First die

Second die	1	2	3	4	5	6
1	2	3	4	5	6	7
2	3	4	5	6	7	8
3	4	5	6	7	8	9
4	5	6	7	8	9	10
5	6	7	8	9	10	11
6	7	8	9	10	11	12

31 The average weight of 5 boys is 70kg. The average weight of 4 girls is 61kg. The average weight of the 9 children in kilograms is

(A) 65 (B) 66
(C) 67 (D) 65.5

32 There are two 3-digit numbers between 200 and 300 which are divisible by 6 and 9 and have the units digit greater than the tens digit. They are:

(A) 288 and 270
(B) 216 and 256
(C) 268 and 234
(D) 216 and 234

33 The design shown is to be coloured so that no two adjoining areas have the same colour. What would be the least number of colours that could be used?

(A) 9 (B) 4
(C) 3 (D) 2

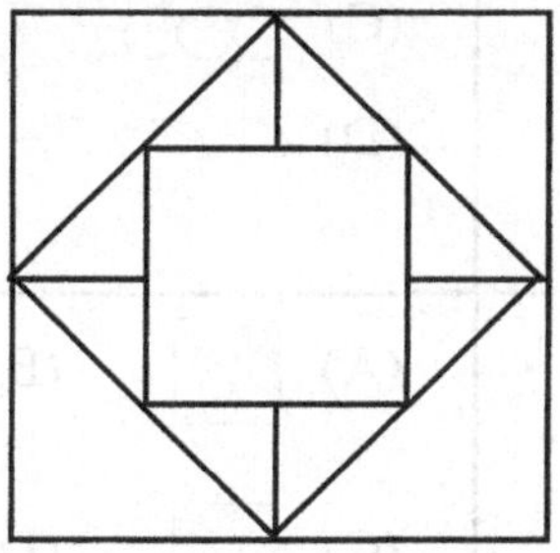

34 The sketch shows the net of a cube. What letter was on the face of the cube opposite the face lettered Q?

		S	U
P	Q	R	
		T	

(A) S (B) U
(C) T (D) R

35 I have a 5 cent, 10 cent, 20 cent and 50 cent coin. What is the total number of amounts of money which I can form from some or all of these coins? (Don't include zero).

(A) 15 (B) 18
(C) 24 (D) 16

36	A party of 18 persons went to a restaurant. Each chose a $21 meal, but 4 of them forgot to bring their money. In order to settle the total bill, those who brought their money each had to pay an extra	(A) $5.25 (B) $4.50 (C) $1.50 (D) $6
37	A chessboard consists of a square made up of 64 small squares of the same size. How many of these small squares are along the perimeter of the chessboard?	(A) 24 (B) 64 (C) 28 (D) 32
38	When doing a series of additions on a calculator, a student noted that she added 35 095 instead of 35.95. In order to obtain the correct total in a single step, she should now	(A) add 35.95 (B) subtract 35 059.05 (C) subtract 35 130.95 (D) add 35 130.95
39	Four students Peter, Quincy, Raymond and Susie are to be seated in a row so that Raymond and Susie are always together. How many different seating arrangements are possible?	(A) 6 (B) 8 (C) 24 (D) 12
40	Consider the numbers 2, 4, 6, ..., 124. How many times does the digit 2 occur in these numbers?	(A) 20 (B) 19 (C) 21 (D) 16
41	X X is a point on a 20c coin. If the coin rolls in a straight line along the top of a flat table, the path of X is:	(A) (B) (C) (D)
42	? What should the fourth square be to obtain a symmetrical pattern of four squares?	(A) (B) (C) (D)
43	Angela is 5 years older than Tony. Tony is 6 years younger than Kate. Angela is 14 years old. How old is Kate?	(A) 25 (B) 3 (C) 13 (D) 15
44	The teacher sat all the students in line. When he numbered them from left to right, Janine was number 6. When he numbered them from right to left, Janine was also number 6. How many students were there?	(A) 10 (B) 12 (C) 11 (D) 13
45	The next number in the sequence 3, 7, 15, 31, ... is:	(A) 65 (B) 53 (C) 63 (D) 73

Selective Schools
Paper 9

1	If H stands for a number and 0.007 x H = 70, then H must stand for	(A) 10 (B) 100 (C) 1 000 (D) 10 000
	Questions 2 - 3: The living expenses for a family are represented by the circle graph (pie chart) shown. Food; Rent 50°; 110°; 80°; Miscellaneous; Clothing	
2	The percentage spent on food is	(A) 25% (B) 35% (C) 66⅔% (D) 33⅓%
3	If the total expenses for a week were $360, how much was spent on food?	(A) $90 (B) $180 (C) $120 (D) $240
4	In the figure given, O is the centre of the circle which has a circumference of 30cm. If the arc ABC = 6cm, then angle AOC is A; B; O; C	(A) 84° (B) 90° (C) 60° (D) 72°
5	A rectangular paddock is of area 0.24 hectares. If the longest side is 60m, then the perimeter, in metres, is	(A) 100 (B) 200 (C) 120 (D) 128
6	247mm as a percentage of 1km is	(A) 2.47% (B) 24.7% (C) 0.0247% (D) 0.247%
7	The value of $\frac{4}{100} + \frac{3}{10}$ is	(A) 0.0043 (B) 0.0034 (C) 0.34 (D) 0.43
8	A girl's average monthly salary for a certain year was $1 010. If her average monthly salary for the first 11 months was $1 000, then her salary for the month of December must have been	(A) $310 (B) $2 010 (C) $1 120 (D) $24 120

9	The difference in the number of axes of symmetry of a rectangle and an equilateral triangle is	(A) 0 (B) 1 (C) 2 (D) 3
10	Two points P and Q are 2 000m apart. The length in centimetres of the join representing PQ on a map drawn to a scale 1 : 25 000 is	(A) $\frac{2}{25}$ (B) 80 (C) $12\frac{1}{2}$ (D) 8
11	An article costs $72.60. If the cost is increased by $\frac{1}{3}$, the new cost is	(A) $48.40 (B) $108.90 (C) $90.75 (D) $96.80
12	The cost of a toy was increased by one fifth. If the toy sold for $30, then the cost price was	(A) $36 (B) $24 (C) $20 (D) $25
13	If you express 0.75 of a kilometre in metres, you get	(A) 75 m (B) 7 500 m (C) 7.5 m (D) 750 m
14	What would be paid in council rates on a property valued at $65 000 at the rate of 2.3c in the dollar?	(A) $149 500 (B) $1 505 (C) $1 495 (D) $1 395
15	The fraction of the rectangle which is shaded is	(A) $\frac{1}{2}$ (B) $\frac{1}{4}$ (C) $\frac{2}{3}$ (D) none of these
16	One tap could fill a bucket in 20 minutes. Another tap could fill the same bucket in 30 minutes. If they were both used to fill the bucket at the same time, it would take	(A) 50 min. (B) 25 min. (C) 12 min. (D) 10 min.
17	15 people are coming to dinner. How many tins of sliced peaches must you buy if each can serves 4 people?	(A) 60 (B) 3.75 (C) 4 (D) 3
18	$77 \div 2 + 9 =$	(A) 11 (B) 47.5 (C) 7 (D) none of these
19	The digits of the number 4 193 are arranged first in ascending order and then in descending order. The difference between these two results would be	(A) 8 082 (B) 7 992 (C) 7 785 (D) 2 790

20 The perimeter of a square is 20cm. The area of the square in cm^2 is	(A) 25 (B) 16 (C) 80 (D) 400
21 The smallest number divisible by 2, 5, 6 is	(A) 12 (B) 60 (C) 30 (D) 24
Questions 22-23 refer to the same rectangle.	
22 The length of a rectangle is 5 times its breadth. If the area of the rectangle is $80cm^2$, then the perimeter of the rectangle is	(A) 48cm (B) 24cm (C) 36cm (D) none of these
23 A square has the same perimeter as the original rectangle. The difference in cm^2, in the areas of the square and the original rectangle is	(A) 25 (B) 64 (C) 0 (D) 36
24 The value of $\frac{0.1 \times 0.4}{5}$ is	(A) 0.08 (B) 0.008 (C) 0.8 (D) 0.0008
25 Which of these nets will <u>not</u> fold into an open box? i) ii) iii) iv) v) vi)	(A) i, iv, v (B) i, ii, iv (C) ii, iii, vi (D) iii, v, vi
26 If July 31st is a Tuesday, what day of the week is July 13th of the same month?	(A) Wednesday (B) Thursday (C) Friday (D) Saturday
27 How many times does the minute hand of a clock rotate during the month of June?	(A) 720 (B) 30 (C) 1 800 (D) 43 200
28 How many 45c stamps may be purchased for $10?	(A) 22 (B) 24 (C) 23 (D) 25
29 It takes Julie 30 minutes travelling at 80km/h to cover the distance from town P to town Q which is due north of P. After completing her business in Q, Julie returns towards P travelling for 20 minutes at 60km/h. At this stage, where will she be from P?	(A) 20km north of P (B) 20km south of P (C) 60km north of P (D) back at P

30	A hemisphere is placed flat side down and cut vertically in halves. What shape is the cut surface?	(A)  (B) 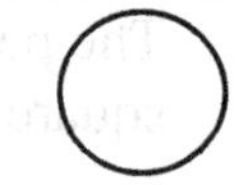 (C) 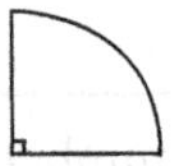 (D) 
31	A certain number between 50 and 70 when divided by 3 leaves a remainder of 1, and when divided by 5 leaves a remainder 3. The number is	(A) 54 (B) 68 (C) 61 (D) 58
32	In the sequence 2, 3, 7, 16, 32, … the next number is	(A) 64 (B) 68 (C) 57 (D) 48
33	If each side of an equilateral triangle is 1cm, find the perimeter of the figure formed by placing 20 such triangles in a row as shown. 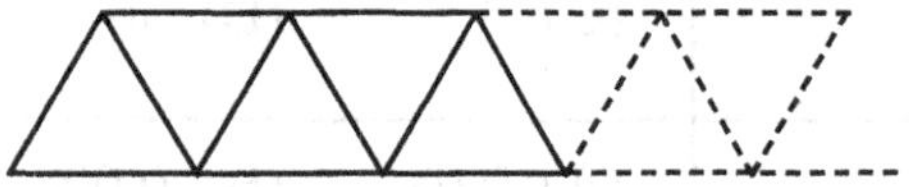	(A) 41cm (B) 60cm (C) 42cm (D) 22cm
34	The numbers on the faces of this cube are consecutive even numbers. What is the largest possible number on the cube? 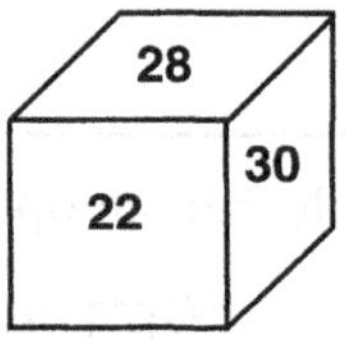 28 30 22	(A) 30 (B) 32 (C) 34 (D) 36
35	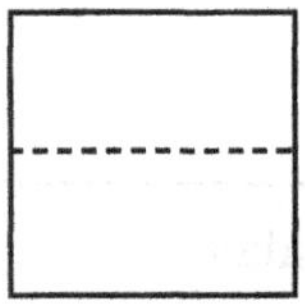 A square sheet of paper is cut into half as shown. If each piece has a perimeter of 18cm, then the perimeter of the original square is:	(A) 24cm (B) 28cm (C) 20cm (D) 32cm
36	Small equal cubes are used to form the first 3 shapes of a pattern as follows: 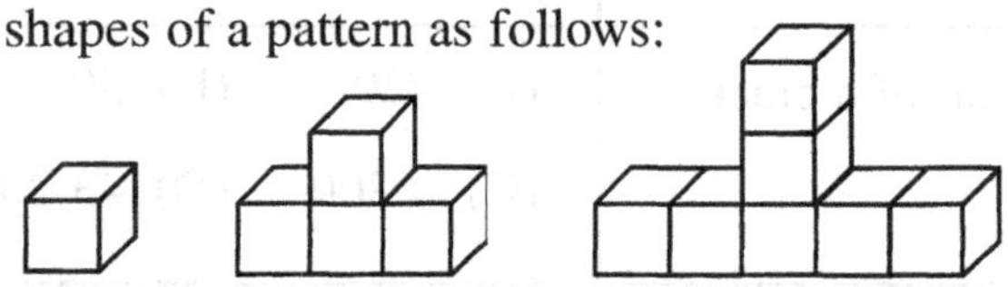 How many small cubes will be used in the 10th shape?	(A) 27 (B) 28 (C) 30 (D) 31
37	The following solids are made of clay: a rectangular prism, a triangular prism, a square pyramid, a triangular pyramid. If each is cut through with a knife which ones could have a triangular section?	(A) triangular prism, square pyramid, triangular pyramid only (B) triangular prism and triangular pyramid only (C) triangular prism and square pyramid only (D) all of them

38	The sketch shows two lengths of the same timber. If Ben can saw the first length into 3 pieces in 3 minutes, how long should it take him to saw the second length into 9 pieces?	(A) 9 min. (B) 12 min. (C) 10 min. (D) 15 min.
39	Which of the following operations with whole numbers will always give a whole number? i) addition ii) multiplication iii) division	(A) i) only (B) ii) only (C) i) and ii) only (D) ii) and iii) only
40	A rectangle is 150cm by 50cm. The area in m^2 is:	(A) 7 500 (B) 75 (C) 7.5 (D) 0.75
41	Which of the given results is approximately equal to 1.97 x 3.142?	(A) 60 (B) 6 (C) 0.6 (D) 0.06
42	Jan and Michelle are given the same amount of pocket money. Jan buys 2 apples and has 70 cents left. Michelle buys 7 apples and has 20 cents left. What amount of money did each receive?	(A) 90c (B) $1.20 (C) 25c (D) $1.00
43	A square of side 10m is to be painted on the concrete quadrangle at a school. By drawing equally spaced vertical and horizontal lines, it is then divided into smaller squares as shown. The total length in metres of all lines to be painted will be:	(A) 100 (B) 80 (C) 60 (D) none of these
44	A block in the shape of a rectangular prism with dimensions 6cm x 3cm x 2cm is carefully dropped into a rectangular tank measuring 8m x 6m x 3m which is full of water. How much water is displaced?	(A) 36L (B) 36mL (C) 144L (D) 108mL
45	Thirty two cubes each of side 1 cm are placed to form four different solids P, Q, R, S as shown in the sketches. P Q R S What is the difference in surface areas between the solids with the greatest and least surface areas?	(A) $10cm^2$ (B) $6cm^2$ (C) $4cm^2$ (D) $12cm^2$

Selective Schools
Paper 10

1	$2^4 + 4^2$ simplifies to	(A) 48 (B) 256 (C) 32 (D) 16
2	If $\frac{3}{5}$ of a number is 42, then the number is	(A) 48 (B) 70 (C) 64 (D) 72
3	Which of the given results has the largest value?	(A) $\frac{3}{5}$ (B) $\frac{1}{2}$ (C) $\frac{9}{20}$ (D) $\frac{3}{4}$
4	If you arrange the numbers 0.41, 0.39, 0.4, 0.49 in ascending order (i.e. from smallest to largest) the correct arrangement would be	(A) 0.49, 0.41, 0.4, 0.39 (B) 0.49, 0.41, 0.39, 0.4 (C) 0.39, 0.41, 0.4, 0.49 (D) 0.39, 0.4, 0.41, 0.49
5	Which of the given numbers is only one tenth as large if you leave out the zero?	(A) 23.10 (B) 2301 (C) 231.0 (D) 2310
6	$\frac{1}{5}$ of 15 is the same as $\frac{1}{4}$ of	(A) 12 (B) 300 (C) $\frac{3}{4}$ (D) $\frac{4}{5}$
7	It takes me 55 minutes to get to school. When should I leave home if I must be there at 9:12 a.m.?	(A) 8:37 a.m. (B) 8:03 a.m. (C) 8:17 a.m. (D) 8:22 a.m.
8	What number must be placed in the boxes to make the number sentence true? ☐ x 8 = 2 x (☐ + 2) + 14	(A) 4 (B) 2 (C) 5 (D) 3
9	Twelve litres of oil can be bought for $16.20. How many litres of oil could you buy for $4.05?	(A) 3 (B) 4 (C) 2 (D) 3.5
10	The most likely number to complete the series 4, 7, 12, 19, 28, ☐ is	(A) 42 (B) 40 (C) 36 (D) 39
11	Concentrated washing powder can be bought in 4 different sizes. Which is the best buy?	(A) 250g for $6.25 (B) 375g for $9.00 (C) 1kg for $25.50 (D) 1.5kg for $37.65
12	If the digits in the hundreds column and the tenths column of the number 312.4 were interchanged, then compared to the original number, the new number would be	(A) smaller by 99.9 (B) larger by 99.9 (C) smaller by 0.1 (D) smaller by 180

13	I went shopping with \$20. Buying school books took one fifth of it. I spent $\frac{3}{4}$ of what was left on a bag. I was left with	(A) \$4 (B) \$5 (C) \$7 (D) \$6.50
14	The sum of $\frac{3}{4}$ and $\frac{1}{2}$ is multiplied by 4. The result is	(A) $2\frac{3}{4}$ (B) $3\frac{1}{2}$ (C) $2\frac{2}{3}$ (D) 5
15	Find half the product of $3\frac{1}{2}$ and 4.	(A) 7 (B) 9 (C) 8 (D) 14
16	One of the given expressions does not equal $\frac{2}{5}$. Which is it?	(A) $\frac{7 \times 2}{5 \times 7}$ (B) $\frac{2+4}{7+8}$ (C) $\frac{2+7}{5+7}$ (D) $(6 + 2) \div (11 + 3 \times 3)$
17	The difference between $4\frac{1}{2}$ and $3\frac{1}{4}$ is	(A) $1\frac{3}{4}$ (B) $1\frac{1}{4}$ (C) $\frac{3}{4}$ (D) $\frac{1}{2}$
18	How many litres of water would be needed to half fill a cubic metre container?	(A) 500 (B) 50 (C) 5 (D) 5 000
19	Each time that a ball bounces, it rebounds to $\frac{3}{4}$ of the height of its previous bounce. If the ball is dropped from a window which is 64m above the ground, how far will it have travelled by the time it hits the ground for the fourth time?	(A) 175m (B) 286m (C) 350m (D) 192m
20	Give the most likely next fraction in the series: $\frac{43}{7}$, $\frac{31}{6}$, $\frac{21}{5}$, $\frac{13}{4}$, $\frac{7}{3}$, □	(A) 1 (B) $\frac{1}{2}$ (C) $\frac{7}{2}$ (D) $\frac{3}{2}$
21	Find the value of 3 x \$21.57 + 2 x \$93.78	(A) \$252.27 (B) \$194.31 (C) \$576.75 (D) \$692.10
22	I bought a bike for \$411.23 but I had to sell it for \$389.76. I lost	(A) \$21.57 (B) \$121.47 (C) \$21.27 (D) \$21.47
23	An estate agent gets 7c in the dollar for collecting rents. One property rents for \$125 per week. For collecting this rent he would receive	(A) \$875 (B) \$87.50 (C) \$8.75 (D) \$0.88

Published by Coroneos Publications

24 The excursion cost each of us \$2.93. There were 200 of us altogether, so we paid a total of	(A) \$5 850 (B) \$493 (C) \$58 600 (D) \$586
25 Find the perimeter of the figure below; (all angles are right angles). 7 12 Units are in cm.	(A) 19 cm (B) 38 cm (C) 84 cm (D) not enough information to tell
26 If X and Y represent digits, then the number 2XY2 is divisible by 8. The values of X and Y must be	(A) X = 3 and Y = 7 (B) X = 8 and Y = 6 (C) X = 9 and Y = 4 (D) X = 8 and Y = 7
27 O is the centre of the circle whose radius is 8cm. A O D C B The total length of AB, OC and OD is:	(A) 16 cm (B) 12 cm (C) 32 cm (D) 24 cm
28 All the angles in the figure are right angles. 4 2 9 11 The area in square units is	(A) 91 (B) 89 (C) 87 (D) not enough information to tell
29 Rounded off to the nearest thousand, the number 147 681 becomes	(A) 147 700 (B) 147 000 (C) 148 000 (D) 148 700
30 2500 divided by 47 is approximately	(A) 53 (B) 50 (C) 60 (D) 49

31 How many square tiles of side $1\frac{1}{3}$cm would be needed to tile a row $9\frac{1}{3}$cm long?

(A) 6 (B) 7
(C) 9 (D) 49

Questions 32 - 34 refer to the following graph.
The graph shows how a girl's salary is spent.

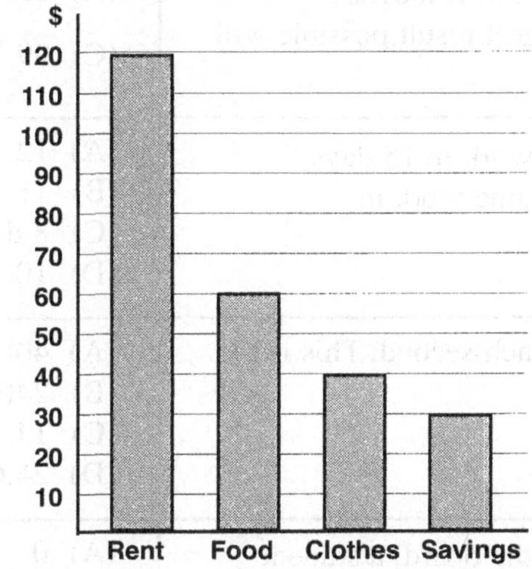

$
120
110
100
90
80
70
60
50
40
30
20
10
Rent Food Clothes Savings

32 What fraction of her salary is spent on rent?

(A) $\frac{1}{2}$ (B) $\frac{3}{25}$
(C) $\frac{12}{25}$ (D) $\frac{3}{4}$

33 What percentage of her salary is saved?

(A) 40% (B) 30%
(C) 12% (D) 25%

34 If her salary were to be put on a pie chart, what angle, to the nearest degree, at the centre of the circle would correspond to the money spent on food?

(A) 86° (B) 60°
(C) 90° (D) 100°

35 324 x 11 = 3 564. If you knew this fact and wanted to use it to find the value of 648 x 33, you would

(A) multiply 648 by 2
(B) multiply 3 564 by 3
(C) multiply 648 by 6
(D) multiply 3 564 by 6

36

2	9	4	12	7
3	17	7	23	

The most likely number which should be placed in the empty box to complete the pattern would be

(A) 13 (B) 33
(C) 83 (D) 271

37	Four of us spent last Saturday collecting for charity. Our collections were: Fred: $23.06 Emma: $57.81 Sam: $12.88 Me: $24.73 What was the average amount that we collected?	(A) $29.62 (B) $31.56 (C) $29.12 (D) $28.98
38	Two whole numbers add to 20. If they are multiplied together, the largest result possible will be	(A) 100 (B) 19 (C) 99 (D) 112
39	If 8 men can do a piece of work in 15 days, then 10 men could do the same work in	(A) 12 days (B) 18¾ days (C) 8 days (D) 10 days
40	A runner covers 4 metres each second. This is the same as	(A) 40km/h (B) 240km/h (C) 14.4km/h (D) 24.6km/h
41	In copying a product from the board, a student copied down 2□7 x 39 = 11 193, but couldn't read the digit where the box is. The digit which should be in the box is	(A) 0 (B) 7 (C) 4 (D) 8
42	Which number should be used to complete the pattern? ? 3 5 8 13 9 25 64	(A) 21 (B) 144 (C) 1 (D) 169

43	Which diagram is the "odd one out"? (A) (B) (C) (D) (E)	(A) (B) (C) (D) (E)
44	How many axes of symmetry does a semicircle have?	(A) 2 (B) 3 (C) 1 (D) as many as you like
45	I am thinking of a number. When I square it and add the result to twice the original number, I get 99. The original number must be	(A) 18 (B) 90 (C) 7 (D) 9

Selective Schools
Paper 11

1	The sum of the first six odd numbers is	(A) 36 (C) 42	(B) 21 (D) 30
2	Which of the given results is the smallest?	(A) $\frac{17}{20}$ (C) $\frac{3}{4}$	(B) 0.8 (D) 70%
3	Subtract the smallest of the following numbers from the largest. 0.10, 0.13, 0.09, 0.11	(A) 0.14 (C) 0.03	(B) 0.04 (D) 0.01
4	$4^3 - 4^2$ simplifies to	(A) 4 (C) 48	(B) 0 (D) 1
5	Find the difference between 239.21 and the number formed by swapping the tens digit and the hundredths digit.	(A) 19.98 (C) 9.9	(B) 99.99 (D) 7.98
6	Four times what number is the same as two thirds of 36?	(A) 6 (C) 48	(B) 96 (D) 12
7	When I passed the starting line, the clock showed 2 minutes 51 seconds and when I passed the finish line it showed 5 minutes 11 seconds. My time for the race was	(A) 2 minutes 20 seconds (B) 2 minutes 40 seconds (C) 3 minutes (D) 3 minutes 40 seconds	
8	What number must be placed in the box so that the number sentence is true? $216 \div 8 = 9 \times \square$	(A) 4 (C) 3	(B) 9 (D) 12
9	On day 1, I started out with $81 and I doubled my money on each day thereafter. Fred started on day 1 with $16 and trebled his money on each successive day. On which day did we have the same amount?	(A) 6 (C) 4	(B) 5 (D) 7
10	You can buy 9kg of cheese for $64.17. How much would 11kg cost?	(A) $52.50 (C) $72.50	(B) $78.43 (D) $69.78
11	Using the digits 4, 1, 9, 7 once only, you can make 24 four-digit numbers. Subtracting the smallest of these from the largest gives	(A) 8 262 (C) 7 722	(B) 6 462 (D) 8 962
12	What is the most likely number to complete the sequence 1, 6, 13, 22, 33, $\square$?	(A) 46 (C) 52	(B) 44 (D) 61

13 You can purchase fruit juices in 4 different sizes. Which is the most economical?	(A) 250mL for $0.34 (B) 600mL for $0.84 (C) 1 litre for $1.35 (D) 2.5 litres for $3.35
14 Half of the sum of ½ and ¾ is	(A) ½ (B) ¼ (C) ⅓ (D) ⅝
15 The perimeter of a circle is called its	(A) radius (B) diameter (C) sector (D) circumference
16 Only one of the following does not equal 10. Which is it?	(A) 2 x 3 + 4 (B) 2 x (6¾ - 1¾) (C) 16 - 2 x 3 (D) $\frac{10 + 10}{2 + 2}$
17 The result of subtracting the numbers 2⅕ and 4³⁄₁₀ is	(A) 2¹⁄₁₀ (B) 1¹⁄₁₀ (C) 2⅖ (D) 3¹⁄₁₀
18 A girl's pocket money is $12.75 each month. How much would she receive in 12 months?	(A) $150 (B) $153 (C) $175.25 (D) $147
19 We bought the ingredients for $7.84 and we sold the lemonade for $43.12. We made	(A) $35.18 (B) $36.28 (C) $35.28 (D) $45.28
20 At the sale I sold 3 games at $19.90 each and 10 toys at $23.81 each. I received	(A) $297.80 (B) $298.80 (C) $835.10 (D) $2 440.70
21 If the tax on furniture is 75%, how much tax would you have to pay on a chair costing $500?	(A) $375 (B) $264 (C) $250 (D) $125
22 $88.65 was shared equally between 9 persons. Each person received	(A) $9.15 (B) $8.95 (C) $9.85 (D) $9.17
23 What is the most likely next number in the series 1, ⅞, ¾, ⅝, ½, □ ?	(A) ⅜ (B) 1 (C) ⅓ (D) ¼

24 The smallest number which is a multiple of the numbers 2, 3, 4, 5 and 6 is	(A) 60 (B) 30 (C) 720 (D) 180
25 Next year, my age added to my brother's age will be 35. The sum of our ages seven years ago would have been	(A) 20 (B) 27 (C) 26 (D) 19

Questions 26 - 27 refer to the following diagram. All the angles are right angles and the diagram is not drawn to scale.

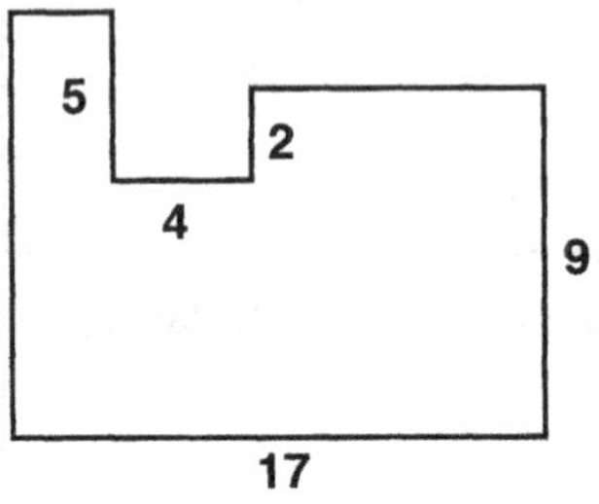

5
2
4
9
17

The units are in cm.

26 The perimeter of the figure is	(A) 57 cm (B) 62 cm (C) 60 cm (D) not enough information to tell
27 The area of the figure is	(A) 160 cm^2 (B) 173 cm^2 (C) 184 cm^2 (D) not enough information to tell
28 One of the given statements is false. Which is it?	(A) All the radii of a circle are equal. (B) The diameter of a circle is the largest chord. (C) An acute angle is smaller than an obtuse angle. (D) An isosceles triangle has three equal sides.

Questions 29 -32 refer to the data in Question 29.

29 The results of a survey of the types of cars that teachers at my school drive were Toyotas 12 Fords 6 Nissans 4 Others 8 What percentage of the teachers drive Fords?	(A) 6% (B) 30% (C) 20% (D) 25%

30 What fraction of the cars are Toyotas or Nissans?

(A) $\frac{8}{15}$ (B) $\frac{1}{3}$
(C) $\frac{2}{5}$ (D) $\frac{1}{2}$

31 If the results of the survey were to be put on a pie chart, the angle at the centre of the sector representing the "Others" would be

(A) 72° (B) 90°
(C) 108° (D) 96°

32 If the results are also to be put on a bar chart of length 15cm, how long in centimetres will the section representing Nissans be?

(A) 2 (B) 8
(C) 6 (D) 4

33 The number 33□2 is divisible by 13. What digit must replace the box?

(A) 5 (B) 4
(C) 0 (D) 1

34 When we drive from Sydney to Grafton at an average speed of 80km/h we take 9 hours. How long, in hours, would we take if we averaged 90km/h?

(A) 10 (B) 8
(C) 7 (D) 11

35 $\frac{636}{12} = 53$. You could use this result to help evaluate $\frac{636}{4}$ by

(A) multiplying 636 by 3
(B) multiplying 53 by 48
(C) multiplying 53 by 3
(D) multiplying 636 by 4

36 5% of a number is 94. The number is

(A) 470 (B) 1 880
(C) $18\frac{4}{5}$ (D) $4\frac{7}{10}$

37 Find the number which will make the number sentence true

720 ÷ □ = 18 x 8

(A) 10 (B) 6
(C) 9 (D) 5

38

3	8	5	4	1	9
10	25	16	13	4	

The most likely number which should be placed in the box to complete the pattern would be

(A) 30 (B) 32
(C) 36 (D) 28

39 I bought 4 shirts.

Shirt 1 cost $39.95
Shirt 2 cost $83.18
Shirt 3 cost $57.88
Shirt 4 cost $63.07

The average cost of the shirts was

(A) $61.02 (B) $72.16
(C) $59.36 (D) $64.08

<table>
<tr><td>40</td><td>Rounded off to the nearest hundred, the number 46 351 becomes</td><td>(A) 46 300 (B) 46 350
(C) 46 400 (D) 46 000</td></tr>
<tr><td>41</td><td>Two four-digit numbers have the same number of hundreds. When you subtract them, the number in the hundreds column of the answer will be</td><td>(A) 0 or 9 (B) 0 or 1
(C) 8 or 9 (D) 1 or 5</td></tr>
<tr><td>42</td><td>In the magic square below, each row, column and diagonal adds to the same number. Find the number which would be in the position marked x.
<table><tr><td>36</td><td></td><td></td><td>30</td></tr><tr><td>14</td><td>26</td><td>24</td><td></td></tr><tr><td>22</td><td></td><td>16</td><td></td></tr><tr><td></td><td>32</td><td>x</td><td>6</td></tr></table></td><td>(A) 34 (B) 31
(C) 28 (D) 20</td></tr>
<tr><td>43</td><td>$2.40 will buy 3kg of flour. Forty cents will buy</td><td>(A) $\frac{1}{6}$kg
(B) $\frac{1}{2}$kg
(C) 800g
(D) none of these</td></tr>
<tr><td>44</td><td>Find the width of a rectangle with perimeter 180cm and length 70cm.</td><td>(A) 70cm (B) 30cm
(C) 20cm (D) 50cm</td></tr>
<tr><td>45</td><td>The diagram shows 12 matches forming 5 squares. What is the least number of matches which must be removed so that only 2 squares remain?</td><td>(A) 1 (B) 2
(C) 5 (D) 4</td></tr>
</table>

Selective Schools
Paper 12

1	$0.05 + \frac{1}{2} =$	(A) 0.1 (B) 0.55 (C) 1.00 (D) 5.5
2	Which given number is closest to 4.18?	(A) 4.2 (B) 4.19 (C) 4.21 (D) 4.165
3	Which number is halfway between 74.81 and 84.37?	(A) 159.18 (B) 9.56 (C) 79 (D) 79.59
4	$5^3 + 5^2$ simplifies to	(A) 390 625 (B) 25 (C) 150 (D) 3 125
5	Which of the given numbers is the smallest?	(A) 23% (B) ¼ (C) 0.24 (D) ⅖
6	Four of us had to stand in line for tickets. There were three girls Ally, Babs, and Candy, and one boy Don. The three girls wanted to stand in line so that they were always together. In how many ways could the four of us stand in line as stated	(A) 6 (B) 12 (C) 16 (D) 8
7	What number must be placed in the boxes to make the number sentence true? 2 x (3 + □) = 13 + □	(A) 8 (B) 7 (C) 10 (D) there can be no answer
8	The tens digit and the hundreds digit in the number 4 917 are interchanged. Compared to the original number, the new number is	(A) smaller by 5 500 (B) larger by 5 500 (C) smaller by 720 (D) larger by 720
9	The boy who came in first in the race took 2min. 47seconds. The second boy took 3min. 12seconds. The first beat the second by	(A) 25seconds (B) 1min. 5seconds (C) 5seconds (D) 35seconds
10	Each week, ⅗ of my salary is spent on rent. That leaves me with $140. How much do I spend on rent?	(A) $196 (B) $210 (C) $56 (D) $350
11	Twice the difference between 3¼ and 1½ is	(A) 3 (B) 2½ (C) 3½ (D) 2

12	5 litres of fuel will take you 35km. How many litres would it take to go 42km?	(A) 7.2 (B) 6 (C) 5.5 (D) 6.5
13	Which of the given quantities of cheese represents the best buy?	(A) 250g for $1.28 (B) 600g for $3.12 (C) 1.2kg for $6.00 (D) 2.5kg for $13.25
14	A rectangular paddock is half as wide as it is long. It is completely enclosed by 3km of wire. Its area in square kilometres must be	(A) 9 (B) 3 (C) $4\frac{1}{2}$ (D) $\frac{1}{2}$
15	How many multiples of 7 are there between 29 and 135?	(A) 14 (B) 15 (C) 16 (D) 13
16	$4\frac{1}{2}$ is divided by 3. The result is multiplied by 2. The answer is	(A) 4 (B) 3 (C) $6\frac{1}{2}$ (D) $\frac{3}{4}$
17	The smallest number that 24 can be multiplied by to give a perfect square is	(A) 24 (B) 12 (C) 6 (D) 18
18	I save $2.25 each week. How much will I have saved in 52 weeks?	(A) $117 (B) $114.75 (C) $127 (D) $123.25
19	Lee bought 8 T-shirts for $2.32 each and sold them for $5.18 each. How much money did she make?	(A) $41.44 (B) $39.12 (C) $22.88 (D) $32.54
20	Each week I save half my pocket money and I buy stamps for my collection with $\frac{1}{5}$ of the pocket money. That leaves $3.21. How much do I spend on stamps?	(A) $2.14 (B) $16.05 (C) $8.03 (D) $32.10
21	I pay tax at an average rate of 30%. How much do I have left after paying tax on my yearly salary of $40 000?	(A) $1 200 (B) $3 200 (C) $12 000 (D) $28 000
22	When we won the competition, the team received $623 which was to be shared equally between the seven players. How much did each player receive?	(A) $89 (B) $91 (C) $99 (D) $79
23	The most likely next number in the sequence 1, $\frac{3}{2}$, $\frac{9}{4}$, $\frac{27}{8}$, ☐ is	(A) $\frac{36}{10}$ (B) 3 (C) $\frac{36}{16}$ (D) $\frac{81}{16}$

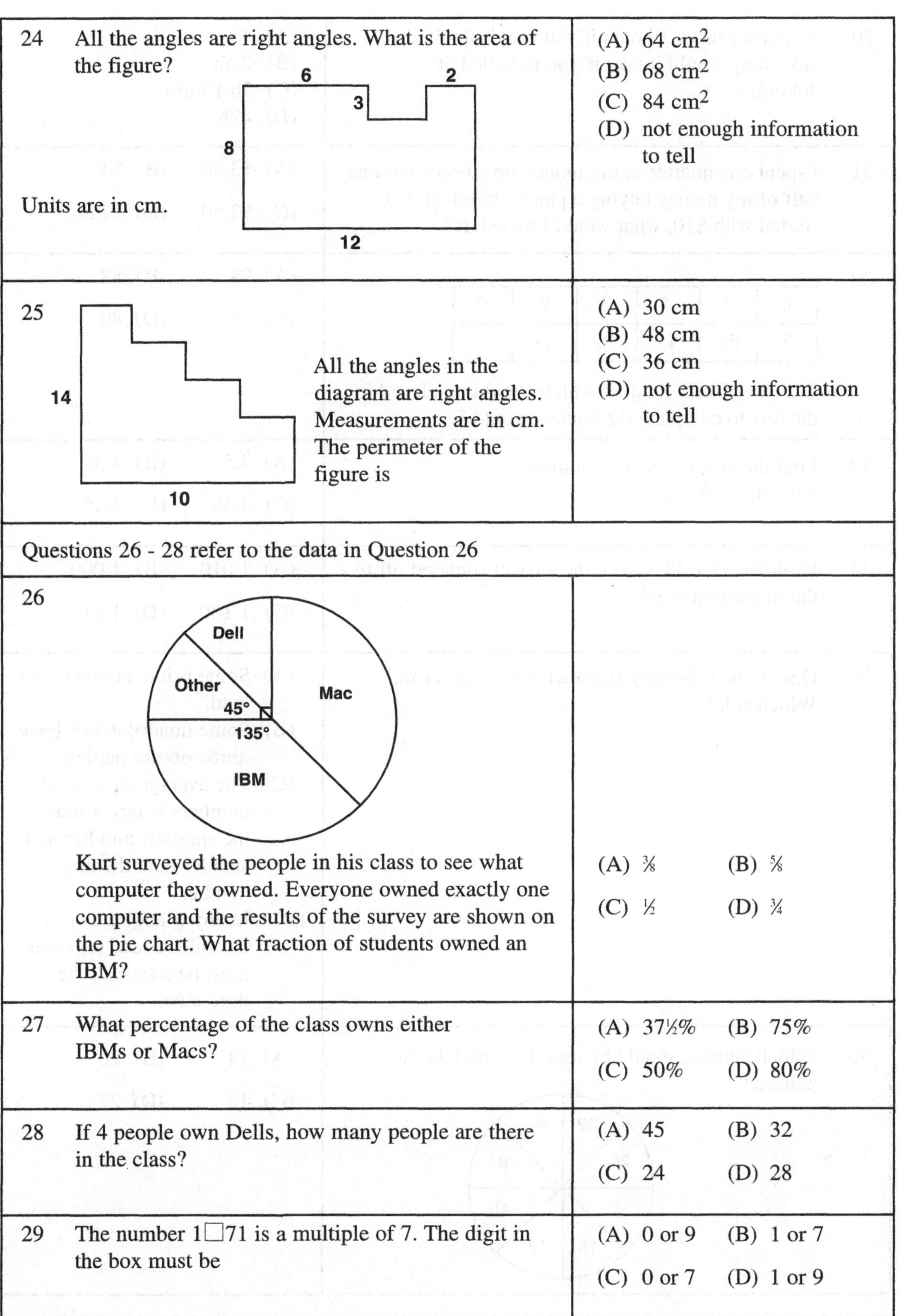

24 All the angles are right angles. What is the area of the figure? 6 2 3 8 12 Units are in cm.	(A) 64 cm^2 (B) 68 cm^2 (C) 84 cm^2 (D) not enough information to tell
25 14 10 All the angles in the diagram are right angles. Measurements are in cm. The perimeter of the figure is	(A) 30 cm (B) 48 cm (C) 36 cm (D) not enough information to tell
Questions 26 - 28 refer to the data in Question 26	
26 Dell, Other, Mac, IBM, 45°, 135° Kurt surveyed the people in his class to see what computer they owned. Everyone owned exactly one computer and the results of the survey are shown on the pie chart. What fraction of students owned an IBM?	(A) $\frac{3}{8}$ (B) $\frac{5}{8}$ (C) $\frac{1}{2}$ (D) $\frac{3}{4}$
27 What percentage of the class owns either IBMs or Macs?	(A) 37½% (B) 75% (C) 50% (D) 80%
28 If 4 people own Dells, how many people are there in the class?	(A) 45 (B) 32 (C) 24 (D) 28
29 The number 1□71 is a multiple of 7. The digit in the box must be	(A) 0 or 9 (B) 1 or 7 (C) 0 or 7 (D) 1 or 9

30 A journey takes 3 hours if you travel at 48km/h. How long would it take if you travelled at 36km/h?

(A) 4h
(B) 2.5h
(C) 2h 15min.
(D) 4½h

31 I spent one quarter of my money on sweets and one half of my money buying a pair of goldfish. If I started with $10, what would I have left?

(A) $3.50 (B) $4
(C) $2.50 (D) $7.50

32

2	7	4	3	6	9
5	50	17	10	37	

The most likely number which should be placed in the box to complete the pattern would be

(A) 94 (B) 82
(C) 78 (D) 80

33 Find the average of the numbers 4.03, 4.1, 4.94, 4.73

(A) 4.5 (B) 4.35
(C) 4.55 (D) 4.45

34 Evaluate 33 x 33, giving the answer rounded off to the nearest hundred.

(A) 1 010 (B) 1 000
(C) 1 100 (D) 1 110

35 One of the following statements must be false. Which is it?

(A) Some prime numbers are odd.
(B) Some quadrilaterals have three obtuse angles.
(C) The average of a set of numbers is larger than the smallest number and smaller than the largest number.
(D) In any triangle, two of the sides added together must be less than the third side.

36 Which number should be used to complete the pattern?

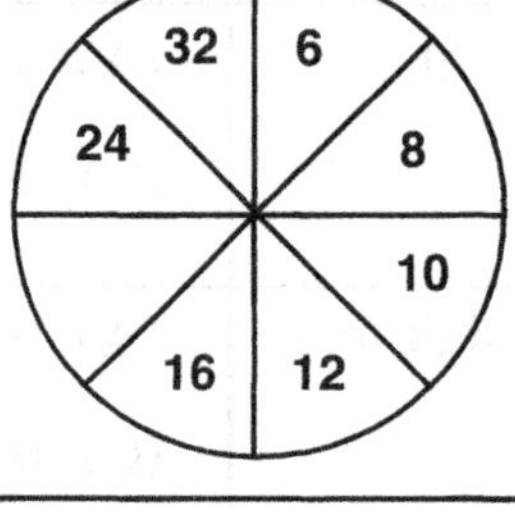

(A) 14 (B) 20
(C) 18 (D) 22

37	Which of the diagrams (A) (B) (C) or (D) would best fit in the space provided in order to complete the pattern? ?	(A) (B) (C) (D)
38	The sketch shows a large square. The midpoints of the sides of the large square are joined to form a smaller square. Which of the given statements is true?	(A) The large square is four times as large as the small square. (B) The large square is twice as large as the small square. (C) The large square is three times as large as the small square. (D) None of these is true.
39	The sum of all the counting numbers from 1 to 50 inclusive, is approximately	(A) 2 500 (B) 1 500 (C) 1 200 (D) 1 000
40	Two whole numbers multiply to give 144. Their sum is 26. The larger of the numbers must exceed the smaller by	(A) 12 (B) 10 (C) 8 (D) 9
41	A girl leaves home and sails 5 nautical miles due south, then west for 3 nautical miles, then north for 2 nautical miles, then east for 10 nautical miles, then south for 9 nautical miles, then west for 7 nautical miles. To reach home she must sail	(A) north for 12 nautical miles. (B) east for 8 nautical miles. (C) north for 16 nautical miles. (D) west for 3 nautical miles.
42	The time at Perth is 2 hours behind the time in Sydney. If it is 11:54 p.m. Monday in Perth, then the time in Sydney would be	(A) 1:54 a.m. Tuesday (B) 9:54 p.m. Monday (C) 1:54 p.m. Tuesday (D) 9:54 a.m. Monday
43	It costs $70 to paint a strip 10cm wide down the centre of a road. To save money it is decided to make the strip 7cm wide. The saving in dollars would be	(A) 12 (B) 21 (C) 49 (D) 19

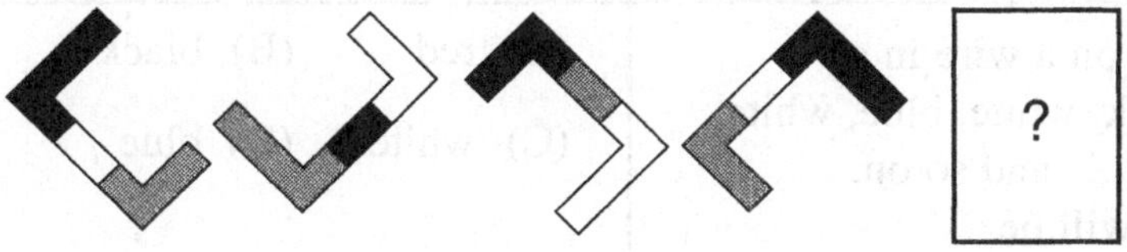

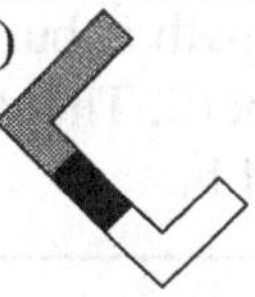

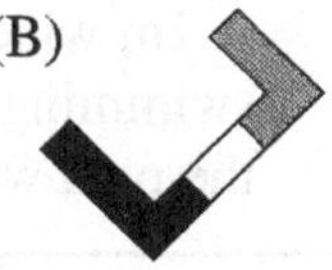

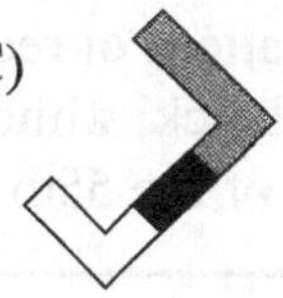

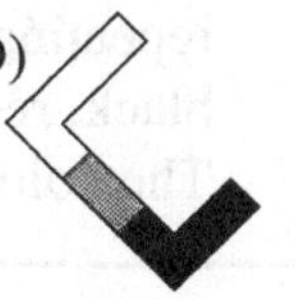

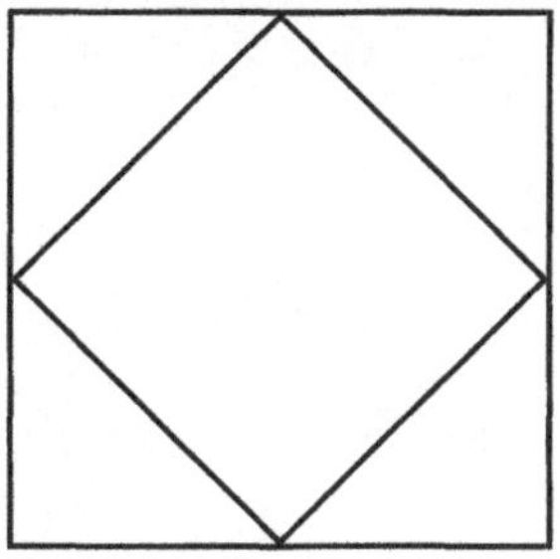

44	A 2m wide path is built around a 20m square swimming pool. The distance around the outside of the path will be	(A) 96m (B) 48m (C) 92m (D) 56m
45	Coloured beads are arranged on a wire in a repeating pattern of red, black, white, blue, white, black, red, black, white, blue,… and so on. The colour of the 55th bead will be	(A) red (B) black (C) white (D) blue

Scholarship
Paper 13

1	There are certain numbers that give no remainder when divided by 3, but when divided by 2 or 5 give a remainder of 1. How many such numbers are there which are less than 100?	(A) 1 (B) 2 (C) 3 (D) 4
2	Uncle Bill is 6 times as old as John and he will be 4 times as old as John in 4 years' time. If Uncle Bill is less than 60 years of age, how old will he be in 6 years' time?	(A) 62 (B) 54 (C) 48 (D) 42
3	A group of seven students buy Christmas presents for each other. How many presents must they buy altogether?	(A) 13 (B) 14 (C) 42 (D) 49
4	Julie was given a mathematics test which contained 25 questions. She answered all 25 questions. She scored 4 marks for each correct answer but she lost one mark for each incorrect answer. If her final mark was 65, how many questions did she get right?	(A) 20 (B) 24 (C) 18 (D) 15
5	I think of a number, multiply it by 6, divide the result by 3, then subtract 5 from that answer. I am left with 75. What was the number I first thought of?	(A) 10 (B) 20 (C) 30 (D) 40
6	How many different ways can the letters W, X, Y and Z be arranged so that the X is the third or fourth letter?	(A) 6 (B) 9 (C) 12 (D) 15
7	Two proof readers are checking the same material. One finds 70 mistakes and the other 94 mistakes. If 39 of the mistakes were found by both readers, how many mistakes were found?	(A) 164 (B) 109 (C) 133 (D) 125
8	George and Ben both decided to swim 10km up a river without leaving the water. Ben swam at an average of 2km per hour. George swam 2km every hour then floated on his back for 10 minutes, but each time he floated the current dragged him back 1km. What was the time difference when they both completed the swim?	(A) 5h 20min. (B) 6h (C) 6h 10 min. (D) 5h 30min.
9	In a concert theatre there is the same number of seats in each row and the rows are straight. My seat is fourteenth from the front and twelfth from the back. It has 7 seats to its left and 9 seats to its right. How many seats are in the theatre?	(A) 400 (B) 442 (C) 416 (D) 425

	Question	Options
10	Four students can sit at each side of a square table. If 30 of these tables are pushed together end-to-end to make one long straight table, how many students can be seated at it?	(A) 120 (B) 360 (C) 248 (D) 240
11	A tiler charged $120 to tile a rectangular floor. At the same rate, how much would he charge to tile a rectangular floor which is three times as wide and twice as long as the original floor?	(A) $720 (B) $360 (C) $240 (D) $180
12	A bus averages 40km/h for 20km and 60km/h for another 20km. The average speed in km/h for the 40km is:	(A) 40 (B) 48 (C) 54 (D) 60
13	The sum of 3 odd numbers is 13. The largest value of their product would be:	(A) 63 (B) 81 (C) 75 (D) 147
14	A fraction which lies between 0 and 1 has both a numerator and denominator. If 5 is added to both the numerator and the denominator, the original fraction has been	(A) unchanged (B) decreased by 5 (C) increased by 5 (D) made closer to 1
15	What percentages of whole numbers from 5 to 24 both inclusive, are exact multiples of 5?	(A) 20% (B) 25% (C) 30% (D) 35%
16	Two runners are running around an oval. They start together and one takes 300 seconds to complete exactly one lap while the other takes 1050 seconds. How many laps would the faster runner make before they meet again at the starting point?	(A) 14 (B) 5 (C) 35 (D) 7
17	There were 9 people at a party. If each person shook hands exactly once with each of the others, how many handshakes were exchanged?	(A) 36 (B) 28 (C) 24 (D) 72
18	In making one revolution a bicycle wheel travels 100cm. How fast are the wheels spinning, in revolutions per minute, when the bicycle is going at 30km/h?	(A) 100 (B) 250 (C) 500 (D) 750
19	A rectangular garden plot 7 metres by 8 metres is to have a concrete path 50 centimetres wide around it. What is the area of the concrete path in square metres?	(A) 7.75 (B) 56 (C) 14 (D) 16

20	On a farm there are only pigs, goats, sheep and cows. Twenty animals are pigs, one tenth are goats, one fifth are sheep and half are cows. How many animals are there on the farm altogether?	(A) 80 (B) 36 (C) 100 (D) 120
21	How many squares are there in the following diagram? 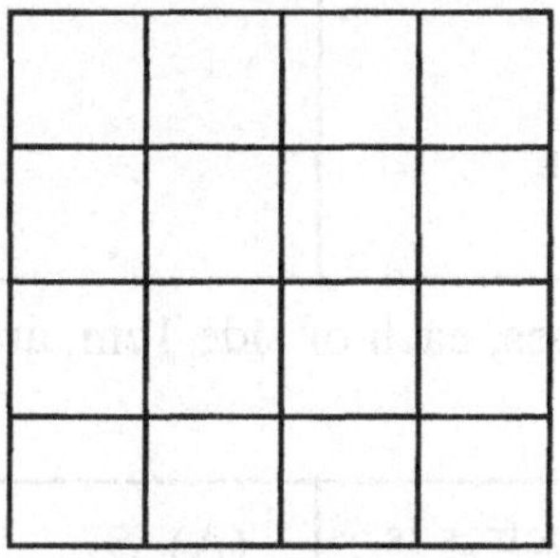	(A) 16 (B) 24 (C) 28 (D) 30
22	How many diagonals can be drawn from any vertex of a figure (on a flat surface) which has 50 sides?	(A) 50 (B) 47 (C) 40 (D) 29
23	What is the starting number? $\square \div 5 \times 4 + 6 - 2.5 = 19.9$	(A) 11.5 (B) 16.05 (C) 20.5 (D) 25.5
24	Four girls were asked to think of a two digit number. Each girl wrote the last digit of her number. Who could have thought of a prime number? (i) Fiona wrote 7 (ii) Liz wrote 5 (iii) Joy wrote 2 (iv) Anne wrote 9	(A) Fiona, Liz (B) Liz, Joy (C) Only Anne (D) Fiona, Anne
25	Three pies and two drinks cost \$8.80; two pies and three drinks cost \$7.20. What would you pay for a pie and a drink?	(A) \$2.40 (B) \$3.20 (C) \$3.60 (D) \$1.80

Scholarship
Paper 14

1 A "strange multiplication" table is shown below:

x	0	1	2	3
0	0	0	0	0
1	0	1	2	3
2	0	2	0	2
3	0	3	2	1

Using it gives 2 x 3 = 2, 3 x 3 = 1 and so on. Which of the given statements is false?

(A) (2 x 3) x 3 = 2
(B) 2 x 3 = 2 x 1
(C) 2 x 2 = 1
(D) 2 ÷ 3 = 2

Questions 2 - 3: One hundred and twenty five cubes, each of side 1cm, are placed on a table to form a solid cube.

2 The length of the edge of the cube in centimetres is

(A) 5 (B) 25
(C) 125 (D) 1

3 The top and the four side faces of this large cube are now painted red. The number of the original 1cm cubes which have just one face painted red is

(A) 36 (B) 57
(C) 24 (D) 54

4 A girl throws 6 dice simultaneously, and writes down the total score. The number of different totals that she could get are

(A) 36
(B) 31
(C) 30
(D) none of these

Questions 5 and 6 refer to the same set of numbers.

5 Two digits are chosen from the digits 1, 2, 3, 4 and 5. Using these two digits a fraction is formed. What is the total number of such fractions which are of different values and are smaller than 1?

(A) 7 (B) 10
(C) 25 (D) 9

6 The average of these fractions is

(A) ⅔
(B) ¾
(C) ½
(D) none of these

Questions 7 - 9: The area of a square is 100cm^2. If the length of each side is decreased by 20%, then the

7 area of the new square in cm^2 is

(A) 144 (B) 120
(C) 64 (D) 80

8 area of the original square has been decreased by

(A) 20% (B) 36%
(C) 64% (D) 80%

9	perimeter of the new square as a fraction of the original perimeter is	(A) $\frac{4}{5}$ (C) $\frac{1}{5}$	(B) $\frac{6}{5}$ (D) $\frac{2}{5}$
10	Which of the given numbers has the greatest number of factors?	(A) 16 (C) 72	(B) 36 (D) 128
	Questions 11-12 refer to the same set of numbers.		
11	The number of counting numbers from 1 to 200 inclusive, which contain the digit 1 at least twice is	(A) 20 (C) 21	(B) 19 (D) 18
12	The number of counting numbers from 1 to 200 inclusive, which contain the digit 1 either once or not at all is	(A) 182 (C) 179	(B) 181 (D) 180
13	A boy buys chocolates at 17c each and some sweets at 5c each. If he spends exactly $1, then the numbers of each that he buys are respectively	(A) 4 and 5 (C) 6 and 2	(B) 3 and 5 (D) 5 and 3
14	On the Reamur temperature scale, water freezes at 0° and boils at 80°, instead of 0° and 100° on the Celsius scale. If the temperature is 24° Reamur, then the temperature in degrees Celsius is	(A) 24 (C) 36	(B) 30 (D) 42
15	When you multiply the ages in years of two teenagers you get 210. The sum of their ages would be	(A) 31 (C) 37	(B) 29 (D) 41
16	Rocky's Primary volleyball team won 5 out of every 7 games they played last season. They lost 10 games. How many games did they play?	(A) 35 (C) 70	(B) 50 (D) 350
17	Alan decided to take a trip to the Gold Coast for a package price of $541.64. The price included airfare and 8 days and 7 nights in a hotel. If the airfare alone was $160, what was the cost of the hotel accommodation per night?	(A) $77.38 (C) $67.71	(B) $47.71 (D) $54.52
18	Jack is the best mathematician in his class. Once he was asked to add all the whole numbers from 1 to 100 inclusive. He had the answer in a couple of minutes. What was Jack's answer?	(A) 200 (C) 5 000	(B) 10 000 (D) 5 050

19	The school flag is being designed. It is decided that there must be 3 vertical coloured strips on it as shown below. Strips next to one another must be different colours. There are 4 different colours to choose from. How many different flags are possible?	(A) 48 (B) 18 (C) 24 (D) 36
20	My father invested his money in some stock. For every \$4.50 he invested, he got back \$9.50. If he ended up with a total of \$228, how much did he invest?	(A) \$45.60 (B) \$57 (C) \$108 (D) \$223
21	On their trip to Sydney, the football team drove 170km in 3 hours 15 minutes, stopped for 40 minutes and then drove 240km in 4hours 20 minutes. What was their average speed for the total time of the trip? Round the answer off to the nearest kilometre per hour.	(A) 50 (B) 65 (C) 55 (D) 45
22	The school photographer had to arrange the students P, Q, R and S in order from tallest to shortest. P is shorter than Q, R is taller than S and S is shorter than P. S ended up being 2 students away from R. How did the photographer arrange the students?	(A) RQPS (B) QRPS (C) QPRS (D) PQSR
23	Mike and Carol like to ride bicycles. Mike can ride 6km in the same time that Carol can ride 4km. They want to ride 18km and finish together. How much of a start must Carol have?	(A) 8km (B) 6km (C) 12km (D) 16km
24	Tom has 5 times as many comic books as Deborah. Deborah has 40% of the number of comic books that George has. If the total number of comics owned by the three friends is 204, how many are owned by Tom?	(A) 120 (B) 180 (C) 165 (D) 175
25	Four blocks have an average weight of 1.5kg. The weight of the heaviest could be:	(A) 6kg (B) 1.5kg (C) 8kg (D) 3kg

Scholarship
Paper 15

1	How many different numbers can be made using the numerals 2, 5, 8 if no numeral can be repeated in any number?	(A) 6 (B) 9 (C) 12 (D) 15
2	A candle is alight for 10 minutes per day. It burns 3mm of its length during that time. Before the candle was lighted on 1st June, its length was 4.2cm. What date will it be when the candle burns out?	(A) 12th June (B) 13th June (C) 14th June (D) 15th June
3	Henry wanted to divide a certain number by 4 to get an answer. However he used the calculator incorrectly and multiplied by 4 instead. His answer was 60. The correct answer would have been	(A) 3.75 (B) 15 (C) 4 (D) 12
4	A tennis club has a knockout Singles Tournament in which anyone who is defeated in any match is eliminated. The tournament lasts until one player remains. If 14 players entered, how many matches will be needed before the winner is decided?	(A) 13 (B) 12 (C) 14 (D) 11
5	Two numbers are reciprocals of one another if their product is 1; thus $1\frac{2}{3}$, $\frac{3}{5}$ are reciprocals since $\frac{5}{3} \times \frac{3}{5} = 1$. When a certain number is added to its reciprocal, the answer is $\frac{41}{20}$. The original number must have been	(A) $1\frac{1}{3}$ (B) $\frac{3}{4}$ (C) $1\frac{1}{4}$ (D) $2\frac{1}{2}$
6	The arithmetic mean (average) of a set of 50 numbers is 38. Two numbers of the set, namely 45 and 55, are removed. The arithmetic mean of the remaining set of numbers is	(A) 35.5 (B) 36 (C) 36.5 (D) 37.5
Questions 7 - 8: The letters a and b stand for numbers. Consider the operation * which is such that a * b means "multiply the number one less than a by the number one more than b".		
7	The value of 4 * 6 is	(A) 18 (B) 21 (C) 25 (D) 35
8	If 3 * k = 10, the value of k is	(A) $3\frac{1}{3}$ (B) 4 (C) $4\frac{1}{3}$ (D) 5

<table>
<tr><td colspan="2">Question 9 - 10:

The digital sum of a number is a one digit number obtained by adding its digits. Example: the digital sum of 27 233 is 8 because 2 + 7 + 2 + 3 + 3 = 17 and 1 + 7 = 8.

A natural number greater than 1 is called a perfect number when the sum of its divisors is twice the number. Example: 28 is a perfect number because its divisors are 1, 2, 4, 7, 14, 28 and the sum of these divisors is 56.</td></tr>
<tr><td>9 What is the digital sum of 1905?</td><td>(A) 4 (B) 6
(C) 7 (D) 15</td></tr>
<tr><td>10 What is the smallest perfect number?</td><td>(A) 2 (B) 6
(C) 12 (D) 28</td></tr>
<tr><td>11 What is the LEAST number of triangles that could be fitted together to form a regular hexagon?</td><td>(A) 3 (B) 4
(C) 5 (D) 6</td></tr>
<tr><td>12 Each week, Debra spends 40% of her salary on rent, 25% on food and clothing, and 15% on entertainment. She saves $170 per week. Her weekly salary is</td><td>(A) $850 (B) $680
(C) $920 (D) $760</td></tr>
<tr><td>13 A thin strip of metal of length 2.9m is cut into 3.8cm sections. About how many sections will there be?</td><td>(A) 77 (B) 8
(C) 73 (D) 76</td></tr>
<tr><td>14 A man plants 12 rows of beans, each row being 10m long. The plants in each row are spaced 50cm apart. If the first and last plants in each row are 25cm from the ends of the row, then the total number of plants is</td><td>(A) 252 (B) 112
(C) 220 (D) 240</td></tr>
<tr><td>15 Two masses P, T are attached to a piece of string PQRST, which is 47cm long. The string passes around three small nails Q, R, S so that the lengths QR, RS are equal.
The mass T is 9cm below S whilst the mass P is 7cm below the mass T. What is the length of QR?

S Q R T P</td><td>(A) 15.5cm (B) 16cm
(C) 12cm (D) 11cm</td></tr>
</table>

16	How many three digit numbers can be made using the digits 3, 5, 6 if repetitions are allowed?	(A) 18 (B) 24 (C) 26 (D) 27
17	How many squares and rectangles of any size can be found in the diagram given?	(A) 29 (B) 30 (C) 31 (D) 34
18	A square PQRS has a side of 20m. Ben walks 3 times around the square in an anti-clockwise direction, then 2¾ times around the square in a clockwise direction and finally 6½ times around the square in an anti-clockwise direction. He ended up at S. How far did he walk and where did he start? P Q 20m S R	(A) 960m, R (B) 980m, R (C) 980m, Q (D) 940m, P
19	Factorial 6 is written as 6! and means $6 \times 5 \times 4 \times 3 \times 2 \times 1 = 720$. $\frac{\text{Factorial 7}}{\text{Factorial 6}}$ is written as $\frac{7!}{6!}$ and means $\frac{7 \times 6 \times 5 \times 4 \times 3 \times 2 \times 1}{6 \times 5 \times 4 \times 3 \times 2 \times 1} = 7.$ What would be the value of $\frac{9!}{5!4!} - \frac{8!}{4!4!}$?	(A) 196 (B) 126 (C) 70 (D) 56
20	Three children Ann, Betty and Charles play a game in which the winner is awarded a prize. If they play three separate games altogether, in how many ways can the prizes be won?	(A) 3 (B) 7 (C) 10 (D) 9
21	A rectangular prism is made up of 42 cubes of side 1cm. If the perimeter of the base is 18cm, then the height of the prism in centimetres is	(A) 6 (B) 2 (C) 7 (D) 3

22	A tank contains a certain amount of water. Each month half of the water in the tank evaporates and an extra litre of water is added. After 5 months, after the litre of water is added, there were 4 litres of water in the tank. How many litres were in the tank originally?	(A) 50 (B) 42 (C) 74 (D) 66
23	6cm 7cm 20cm A block of cheese in the shape of a rectangular prism with dimensions 20cm x 7cm x 6cm is cut into 5 smaller blocks by 4 cuts parallel to the ends of the block. By how many cm^2 will the surface area of the original block be increased by these cuts?	(A) 168 (B) 210 (C) 420 (D) 336
24	A jug holds 1.44L of water. This water is poured into an empty rectangular tank measuring 18cm x 10cm x 15cm. What will be the depth then of the water in the tank?	(A) 12cm (B) 8cm (C) 6cm (D) 4cm
25	Sally found the smallest number which could be divided by each of the numbers from 1 to 10 inclusive. When she divided it by 11, she found that she had a remainder of	(A) 1 (B) 3 (C) 8 (D) 9

Scholarship
Paper 16

1	How many odd numbers less than 60 are exactly divisible by 3?	(A) 9 (B) 10 (C) 18 (D) 27
2	A rectangle has length 24cm and breadth 6cm. The length of the side of a square with the same area as this rectangle is	(A) 16cm (B) 12cm (C) 9cm (D) 18cm
3	Four swimmers P, Q, R and S all start a race together. R can swim faster than S. P can swim faster than Q. S can swim faster than Q. R can swim faster than P. Which swimmer should win the race?	(A) P (B) Q (C) R (D) S
4	Car X travels for ½ an hour at a certain speed, whilst car Y travels for 1 hour at half the speed of car X. Car X travels	(A) four times as far as car Y. (B) twice as far as car Y. (C) half as far as car Y. (D) the same distance as car Y.
5	Five square pieces of cardboard each have side 1 cm. A pattern is formed using the five pieces by placing each piece flat on a table so that each piece <u>must touch</u> at least one other piece of cardboard. The pattern with the largest perimeter has perimeter	(A) 12cm (B) 10cm (C) 20cm (D) 24cm
6	A girl has twice as many 20c coins as she has 5c coins. If she changed all the 5c coins for 10c coins, she would then have	(A) two thirds as many coins as before. (B) five sixths as many coins as before. (C) three quarters as many coins as before. (D) four fifths as many coins as before.
7	If t and u represent numbers, then the statement t - u = u - t is true	(A) only if t = u (B) only if t = 0 or u = 0 (C) for all possible values of t or u (D) for no values of t or u
8	A jar full of jam weighs 541g. Half full of jam, it weighs 315g. The empty jar must weigh	(A) 232g (B) 116g (C) 89g (D) 95g
9	A woman sees the hands of a clock in the mirror, but she cannot see the numbers on the clock face. If the time appears to be 6:45, the time shown on the clock face is	(A) 5:15 (B) 3:25 (C) 6:15 (D) 6:45

Questions 10 - 11: A farmer fences a rectangular area 150m long and 120m wide. The fence posts are spaced 10m apart with a post at each corner.

10	How many posts are there along one of the longer sides of the fence?	(A) 14 (C) 16	(B) 15 (D) 17
11	How many posts are there altogether in the fence?	(A) 27 (C) 108	(B) 54 (D) 60
12	A boy has 5 blue socks, 4 red socks, 7 yellow socks and 1 odd green sock all mixed up in a drawer. If he cannot see the socks, what is the greatest number of socks he should take out of the drawer to be certain of getting a pair the same colour?	(A) 18 (C) 2	(B) 4 (D) 5
13	A bathtub will empty at a uniform rate in 15 minutes. With the plug in, it will fill at a uniform rate in 12 minutes. How long in minutes will it take to fill, if the plug is removed and the tap turned on?	(A) 30 (C) 80	(B) 60 (D) 90
14	If n stands for a whole number, which of the given numbers <u>must</u> be odd?	(A) 3 x n (C) n^2	(B) 2 x n + 1 (D) n x n x n
15	One bell rings every 10 minutes, another rings every 12 minutes. If the bells have just rung together, the time, in minutes, until they next ring together is	(A) 22 (C) 60	(B) 30 (D) 72
16	Which of the given possibilities is the best approximation for $\frac{12.3 \times 4.8 + 16.4}{5.9 \times 3.6}$	(A) 3 (C) 5	(B) 4 (D) 8
17	Given a two digit number, a new three digit number is made from it by putting the digit 5 after it. The new number is then equal to	(A) the old number plus 5. (B) ten times the old number plus 5. (C) 100 plus the old number. (D) 100 times the old number plus 5.	

Questions 18 -23 refer to the diagram below.

A 9cm cube is painted blue.

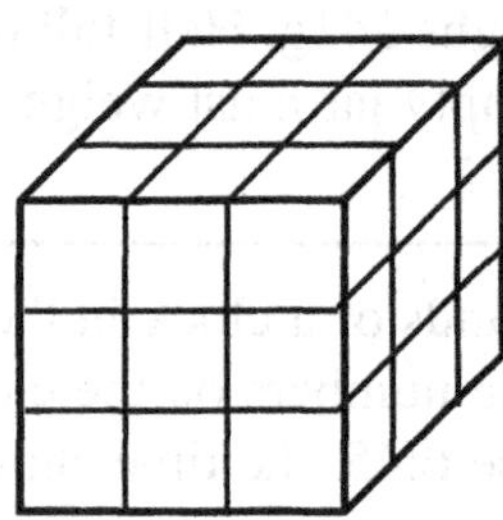

18	The minimum number of cuts needed to cut it into 3cm cubes is	(A) 9 (B) 27 (C) 6 (D) none of these
19	The number of 3cm cubes which are painted blue on 3 sides is	(A) 9 (B) 8 (C) 3 (D) none of these
20	The number of 3cm cubes which are painted blue on 1 side is	(A) 4 (B) 6 (C) 8 (D) 10
21	The number of 3cm cubes which are painted blue on no sides is	(A) 2 (B) 1 (C) 3 (D) 9
22	The number of 3cm cubes which are painted blue on 2 sides as a fraction of the number of 3cm cubes is	(A) $\frac{3}{5}$ (B) $\frac{9}{16}$ (C) $\frac{4}{9}$ (D) $\frac{8}{27}$
23	The number of cubes of sides 3cm or 6cm or 9cm is	(A) 3 (B) 36 (C) 27 (D) 64
24	Arthur is as much taller than Betty as Betty is taller than Charles. Arthur is 1.28m tall. Betty is 1.12m tall. Charles's height is	(A) 1.06m (B) 0.96m (C) 0.16m (D) 1.44m
25	Jim lives in town A and his friend Suzie lives in town B 240 km away. One Saturday he decides to visit her and drives from A to B at an average speed of 80 km/h. He finds Suzie away and returns immediately to A at an average speed of 60 km/h. What is his average speed for the whole trip?	(A) 70 km/h (B) 34 $\frac{2}{7}$ km/h (C) 68 $\frac{4}{7}$ km/h (D) 72 km/h

Scholarship
Paper 17

1	Twenty students in a class play football and the remaining ten play tennis. How many must change from tennis to football so that the percentage of the class which plays football is approximately 83%?	(A) 6 (B) 8 (C) 4 (D) 5
2	When you square a certain number, the answer ends in a 6. The original number must end in	(A) 2 or 3 (B) 4 or 6 (C) 3 or 4 (D) not enough information
3	Tom, Dick and Harry receive a certain sum of money each. Dick gets half of what Tom gets and Harry gets 50% more than Dick. If Dick gets $30, then the amount that they receive altogether is	(A) $135 (B) $75 (C) $105 (D) $100
4	Which of the given fractions lies between $\frac{3}{4}$ and $\frac{3}{5}$?	(A) $\frac{12}{25}$ (B) $\frac{17}{20}$ (C) $\frac{8}{15}$ (D) $\frac{2}{3}$
5	A naval lighthouse consists of 5 lights, one above the other, one or more of which must be on. How many different signals can be shown by the lighthouse? (One of these possible signals is shown in the diagram).	(A) 32 (B) 10 (C) 25 (D) 31
6	A merchant increased the price of a $50 skirt by 10%, but found that he couldn't sell it at that price. So he reduced it from that price by 10%. The skirt was then priced at	(A) $45 (B) $50 (C) $49.50 (D) $55
7	The volume of a cube is 1m^3. The volume of a second cube is 8m^3. The side of the second cube is	(A) twice as large as the side of the first. (B) eight times as large as the side of the first. (C) four times as large as the side of the first. (D) three times as large as the side of the first.

8	In the sketch below all angles are right angles and all measurements are in centimetres. The area of the figure in cm^2 is 7, 9, 11, 3	(A) 78 (B) 69 (C) 71 (D) not enough information to tell
9	In the rectangle below, the measurements are in metres. The shaded area, in m^2 is 8, 12	(A) 60 (B) 96 (C) 48 (D) 72
10	3 x 2½ is the same as ¼ of what number?	(A) 24 (B) 22 (C) 30 (D) 26
11	How many minutes between 11:47 a.m. and 12:36 p.m.?	(A) 49 (B) 83 (C) 43 (D) 57
12	I spent one third of my pocket money, but then my friend gave me $2. I then had $6. My pocket money was	(A) $6 (B) $12 (C) $10 (D) $9
13	This year I paid ⅖ of my salary in tax. That left me with $17 640. How much was my salary before tax?	(A) $10 584 (B) $7 056 (C) $44 100 (D) $29 400
14	A man sees a criminal escaping in a car. He sees only part of the number plate of the car: □XA○37 He tells the police that he couldn't see the first letter or the first number of the number plate, but that he is certain that the other letters and numbers are correct. How many cars would the police have to check in order to find the criminal?	(A) 1 000 (B) 260 (C) 510 (D) 26 000
15	The sum of two numbers is 42 and their difference is 8. The product of the numbers is	(A) 336 (B) 425 (C) 524 (D) 496

16	I am thinking of a number. Half the sum of that number and 11 is the same as 3 more than the number. What is the number?	(A) 4 (B) 5 (C) 8 (D) 13
17	Which two numbers on a number line are twice as far from 2 as they are from 5?	(A) 3 and 6 (B) 4 and 8 (C) 4 and 10 (D) 3 and 5
18	A game is played by 10 people. Nine chairs are arranged in a circle and nine people sit on them. 9 1 2 3 4 5 6 7 8 The tenth person starts counting at chair number 1 and counts in a clockwise direction. He taps every fourth seated person on the shoulder and that person must then stand up. The last of the nine persons still seated wins the game. In what position would you sit in order to win the game?	(A) 1 (B) 3 (C) 4 (D) none of these
19	When the two digit number KY is multiplied by itself, the result is the 3 digit number ZLK i.e. (KY) x (KY) = ZLK. The number that K stands for is	(A) 1 or 9 (B) 2 or 6 (C) 3 (D) 1
20	Which of the given shapes could <u>not</u> have been obtained by rotating the shape marked K? K	(A) (B) (C) (D)
21	What is ⅓ of ⅓ divided by ⅓?	(A) ⅙ (B) $\frac{1}{27}$ (C) ⅓ (D) ½

22	A dog weighs 24kg plus one half of its weight. The dog must weigh	(A) 36kg (B) 48kg (C) 32kg (D) 64kg
23	A store advertises a 15% discount sale. I saved $12 on a can of paint. The original price of the can must have been	(A) $80 (B) $180 (C) $85 (D) $88
24	When thirteen thousand, thirteen hundred and thirteen is written in digits, then the fourth digit from the right will be	(A) 3 (B) 1 (C) 0 (D) 4
25	The 5 digit number 98 7□5 is multiplied by 9 and the answer is increased by 3. If the result is 888 888, then the □ must stand for	(A) 3 (B) 4 (C) 5 (D) 6

Scholarship
Paper 18

1	How many different pairs of parallel edges are there in a cube?	(A) 12 (B) 18 (C) 8 (D) 16
2	If □ stands for a number greater than 4, which of the following is the greatest number?	(A) $\frac{2 \times \square - 1}{5}$ (B) $\frac{2 \times \square}{5}$ (C) $\frac{5}{2 \times \square - 1}$ (D) $\frac{5}{2 \times \square - 2}$
3	Exactly 35% of a group of persons said that they agreed with the proposal. What is the smallest number of persons who could have been asked?	(A) 350 (B) 35 (C) 20 (D) 100
4	"The square of a number is less than the number". This statement is true for	(A) all numbers greater than or equal to 1. (B) all numbers between 0 and 1. (C) all numbers greater than zero. (D) none of these.
5	The sum of three consecutive odd numbers is	(A) odd. (B) divisible by 3. (C) odd and divisible by 3. (D) none of these.
6	If a and b are counting numbers and i) a + b is odd and ii) a x b is odd then	(A) a and b are both odd (B) a and b are both even (C) one of the numbers is odd and the other even (D) statements i) and ii) cannot both be true
7	P, Q and R are whole numbers greater than 1. Which of the following COULD be a prime number?	(A) R + R (B) 3 x P x Q (C) P x Q + Q (D) P + 1
8	One side of a rectangle is increased by 10%. The other side is increased by 20%. The area of the rectangle will increase by	(A) 15% (B) 200% (C) 30% (D) 32%

9	Six children buy presents for each other. How many presents must they buy in total?	(A) 6 (B) 30 (C) 18 (D) 24
10	In the diagram below, PR and QS are diameters. P Q 30° S R What fraction of the circle is shaded?	(A) $\frac{5}{6}$ (B) $\frac{3}{4}$ (C) $\frac{2}{3}$ (D) $\frac{3}{5}$
11	A bottle and a cork cost together 13 cents. If the bottle costs 12 cents more than the cork, how much does the bottle cost?	(A) 11 cents (B) 12 cents (C) 10 cents (D) none of these
12	The average age of ten people at a party is 14 years. When two people leave, the average age rises to 15. If both people were the same age, that age in years must be	(A) 20 (B) 15 (C) 10 (D) 12
13	I can drive 720km from Sydney to Queensland in 9 hours. If I left an hour later than usual, how much greater must my average speed be if I must arrive at the same time?	(A) 20km/h (B) 10km/h (C) 15km/h (D) 40km/h
14	An 8 litre can filled with cooking oil weighs 42kg. When 5 litres are emptied from the can, the can and the oil weigh 27kg. What is the weight of the empty can?	(A) 10kg (B) 18kg (C) 12kg (D) 20kg
15	If 24 men can construct a kit home in 18 days, how many days would it take 27 men, working at the same rate?	(A) 16 (B) 15 (C) 12 (D) 10
16	If one square has twice the perimeter of another, how much larger is its area?	(A) twice as large (B) four times as large (C) eight times as large (D) none of these
17	A boy spent two fifths of his money and found that two fifths of the remainder was $1.20. How much did he have at first?	(A) $4.80 (B) $7.50 (C) $5.00 (D) $6.50

18 A bottle is half full of water. When another 450mL of water is added, the bottle is $\frac{2}{3}$ full. What is the maximum amount of water that the bottle will hold?	(A) 2 litres (B) 2 400mL (C) 3 litres (D) 2 700mL
19 At my favourite restaurant, I can choose from four entrees, five main courses and three desserts. How many different three course meals are possible?	(A) 12 (B) 16 (C) 60 (D) 45
20 The diagrams below show two balances. 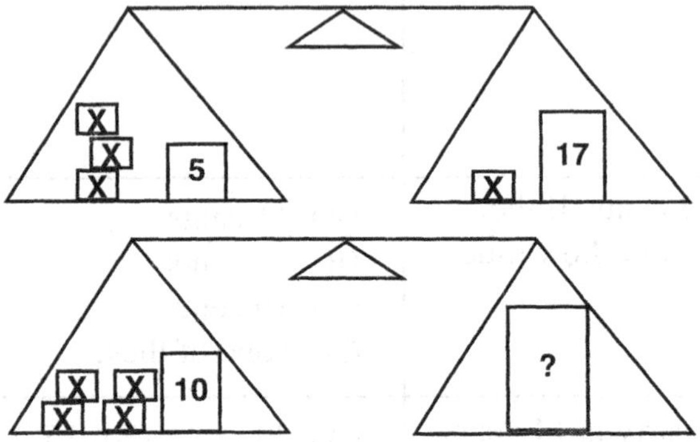 In the two diagrams the boxes marked with an "x" weigh the same. The other boxes are marked with their weight in kilograms. The weight in kilograms of the unknown box (with the ? in it) must be	(A) 26 (B) 34 (C) 30 (D) 36
21 How far in metres will a car travelling at 54km/h go in 12 seconds?	(A) 180 (B) 240 (C) 672 (D) 512
22 In a barnyard, there are only sheep and hens. Altogether there are 37 animals and between them there are 134 legs. How many hens are there?	(A) 15 (B) 20 (C) 12 (D) 7
23 Starting at the same time, two boys walk towards each other keeping in step. If the first boy's step is 70cm long and the second boy's step is 80cm, and they meet after taking 224 steps each, how far apart did they start?	(A) 224m (B) 360m (C) 448m (D) 336m
24 The price of a pen is increased by 20% to $5.82. What was the original price?	(A) $1.16 (B) $4.68 (C) $4.85 (D) $4.66
25 A number is increased by 25% and the result is decreased by 25%. If the final number is 15, the original number was	(A) 15 (B) 12 (C) 20 (D) 16

Scholarship
Paper 19

1	If □ is an unknown number and 6 x □ + 4 =12, then 8 x □ + 5 =	(A) 15 (B) 16 (C) $13\frac{2}{3}$ (D) $15\frac{2}{3}$
2	The 3 in 231 is how many times larger than the 3 in 491.3?	(A) 10 (B) 1 000 (C) 100 (D) 10 000
3	It takes me $\frac{1}{3}$ of an hour driving at 66km/h to cover a certain distance. How fast, in km/h would I have to travel to cover twice this distance in $\frac{1}{2}$ hour?	(A) 99 (B) 44 (C) 396 (D) 88
4	The next number in the series 341, 213, 149, 117, 101 is probably	(A) 83 (B) 95 (C) 99 (D) 93
5	Everybody in a class plays either football or cricket. Two-fifths of the class plays football. Of those that play football, half also play cricket. If 12 students play cricket, the number of people in the class is	(A) 30 (B) 60 (C) 15 (D) 20
6	A sheet of paper is 0.30mm thick. The sheet is folded once, then again and so on. How thick would the folded paper be after 7 folds?	(A) 3.84cm (B) 2.1mm (C) 19.2mm (D) 2.1cm
7	Each card in a deck of playing cards measures 4.5cm by 8.5cm. The total surface area - including the front and back of each card, in a pack of 52 cards is	(A) 3 978cm^2 (B) 1 989cm^2 (C) 38.25cm^2 (D) 76.5cm^2
8	My bathroom scales show that I weigh 10% more than I actually do. What would my weight be if they show me weighing 92.4kg?	(A) 83.16kg (B) 82.4kg (C) 102.4kg (D) 84kg
9	Six men can plough a field in 5 days. How many days should it take 4 men working at the same rate?	(A) $3\frac{1}{3}$ (B) 8 (C) $2\frac{1}{2}$ (D) $7\frac{1}{2}$
10	The surface area of a cube is 216m^2. What is its volume in cubic metres?	(A) 216 (B) 36 (C) 108 (D) 1 024
11	$18 is divided between three children so that the first gets $\frac{1}{2}$ of it, the second gets $\frac{1}{3}$ of it, and the third gets $\frac{1}{6}$ of it. If any of the $18 is left over it is to be given to the child who receives least. That child will receive	(A) $2 (B) $3 (C) $3.50 (D) $4

12	Code words of 1 letter, 2 letters and 3 letters are to be made using the letters A, B, C. How many code words can be made if no letter can be used twice in any code word?	(A) 27 (B) 9 (C) 12 (D) 15
13	Alan has 14 coins in his pockets. They are 5c, 10c and 20c coins. The total value of the coins is $1.75. He has twice as many 10c coins as 5c coins. How many 20c coins are there?	(A) 8 (B) 2 (C) 5 (D) 11
14	I think of a number and square it. From the answer I subtract the original number. The result could NOT be	(A) 30 (B) 42 (C) 132 (D) 18
15	How many different fractions more than 1 can be formed using the digits 2,3,4,5,6 once only? (Each fraction may use only 2 digits).	(A) 11 (B) 10 (C) 9 (D) 8
16	The sum of three consecutive odd numbers could NOT be	(A) 345 (B) 255 (C) 237 (D) 383
17	A centimetre of rain falls on a hectare of land. What would this rain weigh in tonnes?	(A) 10 (B) 100 (C) 10 000 (D) 1 000
18	The length of a rectangle is 1½ times its breadth. If its area is 96cm², its perimeter in centimetres must be	(A) 40 (B) 24 (C) 48 (D) 36
19	The symbol 4! means 4 x 3 x 2 x 1. Using this symbol, the value of $\frac{5! + 6!}{4!}$ would be	(A) 1 663 200 (B) 150 (C) 35 (D) 725
20	T is the midpoint of the side PS of the rectangle PQRS. PS = 2cm and U is on SR so that SU = 2cm, UR = 1cm. What fraction of the rectangle is the shaded area?	(A) ⅔ (B) ¾ (C) ⅚ (D) ⅞

21	'Every even number can be written as the sum of two prime numbers', e.g. 10 = 3 + 7 = 5 + 5. Noting 1 is not considered to be a prime number, in how many ways can the number 50 be written as the sum of two prime numbers?	(A) 4 (B) 5 (C) 6 (D) 7
22	For all numbers represented by the letters a, b the operation * means $a*b = a^2 - a \div b$. The value of 6*12 is	(A) 11.5 (B) 35.5 (C) 36.5 (D) 2.5
23	How many days are there *between* 24 February 2000 and 11 May 2000?	(A) 76 (B) 75 (C) 78 (D) 77
24	How many square centimetres are there in the surface area of a solid cube of side 1 metre?	(A) 600 (B) 6 000 (C) 60 000 (D) 600 000
25	The sketch shows a square tray of area 400cm², in which 16 equal sized coins are placed. What is the radius of each coin?	(A) 5cm (B) 4cm (C) 2cm (D) 2.5cm

Scholarship
Paper 20

1	25% of 40% is equivalent to	(A) 15% (B) 75% (C) 65% (D) 10%
2	Find the difference between the largest of the following fractions and the smallest. $\frac{6}{25}$, $\frac{3}{10}$, $\frac{1}{4}$, $\frac{1}{5}$	(A) $\frac{1}{100}$ (B) $\frac{3}{50}$ (C) $\frac{1}{20}$ (D) $\frac{1}{10}$
3	The next number in the series 11, 15, 24, 40, 65 is probably	(A) 91 (B) 104 (C) 94 (D) 101
4	Max estimates that there are 20 to 40 golf balls in a box. When he counts the balls by fives there is one ball left over. When he counts them by threes there are two left over. What is the exact number of golf balls in the box?	(A) 26 (B) 21 (C) 24 (D) 36
5	A cube of side 6cm has a square hole of side 4cm cut through it. The remaining volume in cubic centimetres is	(A) 96 (B) 64 (C) 120 (D) 152
6	A ladder 12m long rests against a wall. If the foot of the ladder is 6m away from the wall, then the ladder reaches a certain distance up the wall. The best approximation to this distance is	(A) 6m (B) 12m (C) 10m (D) 13m
7	Barry has $7 more than Saul, and Saul has $19 less than Alla. If they have $32 altogether, Saul must have	(A) $24 (B) $5 (C) $2 (D) $11
8	A man has twin sons of equal height. The combined height of the man and one son is 180cm. The combined height of the man and both sons is 252cm. Twice the man's height added to the height of one son would be	(A) 288cm (B) 216cm (C) 324cm (D) 296cm
9	An old way of measuring temperature was the "Fahrenheit" scale. To change a measurement in Fahrenheit degrees into Celsius, you multiplied the number of Fahrenheit degrees by 5, subtracted 160 from your answer, and divided that answer by 9. What would 40°Celsius be in Fahrenheit degrees?	(A) $4\frac{4}{9}$ (B) $779\frac{7}{9}$ (C) 104 (D) 86
10	The first digit of a four digit number is 4 and the last digit is 7. If the first and last digits are swapped, then relative to the old number, the new number would be	(A) larger by 2 997 (B) smaller by 2 997 (C) larger by 3 003 (D) not enough information to tell

11 The shaded figure below was cut from a square piece of metal with a side of 6m. All measurements are in metres and all the angles are right angles. The figure is not drawn to scale. The area in square metres of the figure would be:

(A) 14
(B) 16
(C) 28
(D) not enough information to tell

12 In a class there are 64 students. Ten do not play any sport, 33 play cricket, and 28 play tennis. The number of students who play both cricket and tennis is

(A) 7
(B) 3
(C) 5
(D) none of these

13 The denominator of a fraction is greater than the numerator. If the numerator and denominator of the fraction are both increased by the same amount, then the new fraction

(A) is equal to the old fraction.
(B) exceeds the old fraction.
(C) is smaller than the old fraction.
(D) is greater than or equal to the old fraction depending on the size of the number which was added to the numerator and denominator.

14 In class P, there are 20 students and in class Q there are 28 students. Both classes sit for the same test. The average of the students in class P was 60% and the average of the students in class Q was 70%. It was then decided to put all the marks together and find the common average. This average to the

(A) 64% (B) 65%
(C) 66% (D) 55%

nearest % would be

15 Two weighings are done on a balance. The boxes on the left hand side of the balances all weigh the same and the weight of each ball on the left hand side of the balances is the same.

19kg

21kg

The weight of three boxes and four balls in kilograms would be

(A) 20 (B) 25

(C) 26 (D) 27

16 There are 4 boys P, Q, R, S.
If P is taller than R,
and Q is taller than S,
and P is shorter than Q
and R is shorter than S,
then if we arrange the boys in order of height from smallest to largest, only one of the following is true.

(A) R is shorter than S who is shorter than P who is shorter than Q.
(B) R is shorter than S who is shorter than Q who is shorter than P.
(C) R is shorter than P who is shorter than Q who is shorter than S.
(D) there is not enough information to tell.

17 Water flows into a 48 litre tank at a rate of 3 litres every 15 seconds. At the same time, water flows out of the tank through a tap at a rate of 3 litres each 10 seconds. If the tank was half full to start with, after how many minutes will it be empty?

(A) 8 (B) 12

(C) 2 (D) 4

Questions 18 - 25 apply to the following tables.

In a strange arithmetic, addition and multiplication are worked out using the following tables:

+	0	1	2	3	4
0	0	1	2	3	4
1	1	2	3	4	0
2	2	3	4	0	1
3	3	4	0	1	2
4	4	0	1	2	3

X	0	1	2	3	4
0	0	0	0	0	0
1	0	1	2	3	4
2	0	2	4	1	3
3	0	3	1	4	2
4	0	4	3	2	1

18 Use the tables to find the value of 3 + 2 + 3 x 2.	(A) 0 (B) 1 (C) 2 (D) 3
19 Use the tables to find the value of 3^2 + 2x(2 + 4 x 3).	(A) 4 (B) 37 (C) 0 (D) 2
20 Use the multiplication table to find the value of the cube of 2 i.e. 2 x 2 x 2.	(A) 4 (B) 3 (C) 2 (D) 1
21 Use the multiplication table to find the value of the square root of 4.	(A) 2 only (B) 2 or 3 (C) 3 only (D) 1
22 Using the addition table, find the value of 2 - 3.	(A) 1 (B) 2 (C) 3 (D) 4
23 Use the multiplication table to find the value 2 ÷ 3.	(A) 1 (B) 2 (C) 3 (D) 4
24 If K stands for a number, and 4 + K = 0, use the addition table to find what number K must stand for.	(A) 4 (B) 1 (C) 3 (D) 2
25 If P stands for a number, and 3 x P + 2 = 4, use the tables to find what number P must stand for.	(A) 1 (B) 2 (C) 3 (D) 4

Solutions

PAPER 1

1 (D) Sum of the 6 numbers is 6 x 4 = 24; sum of the 7 numbers is 7 x 5 = 35. Seventh number is (35 - 24) = 11.

2 (C) A millenium (plural millenia) is 1 000 years.

3 (A) Note 5 x $2\frac{1}{2}$ = $12\frac{1}{2}$

4 (C) Note 0.9 = 0.90

5 (B) Multiples of 3 are 3, 6, 9, <u>12</u>, 15, …; multiples of 4 are 4, 8, <u>12</u>, 16, …; multiples of 6 are 6, <u>12</u>, 18, …; Lowest common multiple (underlined) is 12.

6 (C) Note $1\frac{1}{4}$ = 1.25

7 (D) $\frac{1}{2} = \frac{5}{10}$; $1\frac{5}{10} + 2\frac{4}{10} = 3 + \frac{5+4}{10} = 3\frac{9}{10}$

8 (D) \$299.99 is between \$299.00 and \$300.00 and nearer to \$300.00

9 (A) Note freezing point of water is 0°C

10 (C) **Method 1** By trial, find one tenth of each possible answer and multiply it by 3, to see if you obtain 120.

Method 2
If $\frac{3}{10}$ of the number is 120
then $\frac{1}{10}$ of the number is 120 ÷ 3 = 40
thus $\frac{10}{10}$ of the number i.e. the number itself is 40 x 10 = 400.

11 (B) Easy to see answer if you 'make it simpler', thus in 890 there are 8 hundreds and in 7 890 there are 78 hundreds and so on.

12 (D) average = $\frac{3.1 + 8.1 + 5.1 + 7.1 + 3.1 + 4.1}{6} = \frac{30.6}{6} = 5.1$

13 (A) V = L x B x H = 50 x 40 x 30 = 60 000cm^3
But 1cm^3 = 1 mL,
thus 60 000cm^3 = 60 000mL = 60L.

14 (B) In 219.07, the place value of the 2 is 2 hundreds, 1 is 1 ten, 9 is 9 units, 0 is 0 tenths, 7 is 7 hundredths.

15 (B) Note 5 x $1\frac{1}{5}$ = 5 x 1 + 5 x $\frac{1}{5}$ = 5 + 1 = 6.

16 (C) Each week he loses 200g; thus after 50 weeks he will have lost 50 x 200 = 10 000g = 10kg.

17 (D) If 24kg cost \$28.80
then 1kg costs \$28.80 ÷ 24 = \$1.20
thus \$3.60 will buy you 3kg flour

18 (C) Easiest done by addition

$$\begin{array}{r} 8\,000\,000 \\ 800 \\ 8\,+ \\ \hline \end{array}$$

19 (C) There are 360° in a circle. One quadrant (fourth) is equal to 90°; half a quadrant contains 45°.

20 (B) Multiplication and division are always done before addition or subtraction. When we have x and ÷ we work from left to right, i.e. in 36 ÷ 4 x 7 we first calculate 36 ÷ 4 (and then multiply by 7).

21 (D) Daily rate per painter = (\$2 265 - \$465) ÷ 12.

22 (C) The difference in the numbers is always 4.

23 (A) Answer is (28 x 3) + (29 + 11) = 84 + 40 = 124.

24 (B) $\frac{26}{50} = \frac{52}{100}$ = 52%

25 (C) 21.87 = 20 + 1 + $\frac{8}{10}$ + $\frac{7}{100}$

26 (C) Descending order is from largest to smallest.

27 (B) Acute angles are less than 90°

28 (D) The first 3 shapes are solids (i.e. 3-dimensional); a triangle is a 'plane' (flat) figure (i.e. 2-dimensional).

29 (C) Note 4 716 ÷ 1 000 = 4.716, 4 716 x 1 000 = 4 716 000,
4 716 + 1 000 = 5 716; 4 716 - 1 000 = 3 716.

30 (C) 12h 59min = 13h - 1min.
11:35 a.m. Monday + 13h - 1min. = (11:35 a.m. Monday + 1h - 1min.) + 12h = 12:34 p.m. Monday + 12h, i.e. 12:34 a.m. Tuesday.

31 (C) If 30 are boys; 70 are girls; 70/100 = 70%

32 (B) 6 184 - 4 186 = 1 998; 6 184 is larger than 4 186 by 1 998.

33 (A) There are 8 lots of 125mL in 1L.

34 (B) A quadrilateral has 2 pairs of opposite sides; if one pair is parallel, it is a trapezium; if two pairs are parallel, it is a parallelogram..

35 (B) The number must have been 1 093 - 784 = 309. Thus she should have obtained 784 - 309 = 475

36 (C)

37 (A) Change each fraction to have a denominator of 100;
thus $\frac{4}{5} = \frac{80}{100}$, $\frac{39}{50} = \frac{78}{100}$, $\frac{19}{25} = \frac{76}{100}$, …, $\frac{7}{10} = \frac{70}{100}$. Missing fraction = $\frac{72}{100} = \frac{18}{25}$

38 (B) All radii are equal; since AB = 4cm, then BC = CD = DE = 4cm.

39 (C) The perimeter does not include the X.

40 (B) Given shape is made up of 5 squares each of area $16cm^2$,
i.e. total area is 5 x 16 = $80cm^2$.
We want one quarter of this, i.e. 80 ÷ 4 = $20cm^2$.

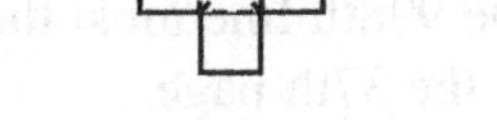

41 (C) You need 4 jugfuls to get 1 litre and thus 4 x 9 = 36 jugfuls to get 9 litres.

42 (D)
$5 A = $16 HK
i.e. $1 A = $16 ÷ 5 HK
thus $64 A = $$\frac{16}{5}$ x 64 HK, i.e. $ $\frac{1024}{5}$ HK
= $$204\frac{4}{5}$ HK, i.e. $204.80 HK

43 (D) No. of edges is 12 and no of vertices is 8.

edge
vertices

44 (A) In decreasing order of size from the largest, the possible numbers are 986 420, 986 402, 986 240, 986 204, 986 042, and so on.

45 (C) She receives 40 000 x 5c + 80 000 x 2c, i.e. $2 000 + $1 600 = $3 600.

PAPER 2

1 (C) Only the pair 11, 17 are both prime, with sum 28 and difference 6.

2 (A) From the largest, in decreasing order, the numbers are 76 521, 76 512, 76 251, …

3 (C) 2 ÷ 1 = 2, 2 x 1 = 2, 2 - 0 = 2

4 **(B)** 4) 298 ; Answer 74 rem 2 = 74¾ = 74½ = 74.5
74 rem 2

5 **(D)** 279 x 30 = 30 x 279, which can be read as 30 lots of 279; similarly 279 x 4 can be read as 4 lots of 279.
30 lots of 279 + 4 lots of 279 = 34 lots of 279
= 34 x 279, i.e. 279 x 34

6 **(C)** 11 wholes = 11, 49 thousandths = $\frac{49}{1000}$ = 0.049; answer = 11 + 0.049 = 11.049

7 **(A)** Note 6.24 ÷ 3 = 2.08, 6.24 x 3 = 18.72, 6.24 - 3 = 3.24, 6.24 + 3.00 = 9.24 (note 3 = 3.00)

8 **(C)** Order from smallest to largest (i.e. ascending order) is 0.56, 0.6 (i.e. 0.60), 0.61, 0.65

9 **(A)** **Method 1** \$40 - \$7.50 = \$32.50; \$32.50 ÷ \$2.50 = 13;
time of call = 3min. + 13min. = 16 minutes

Method 2 Try each of the possible answers
Thus 16min. = 3 min. + 13min.
Cost 16min. call= \$7.50 + 13 x \$2.50 = \$7.50 + \$32.50, so time is 16 min.

10 **(D)** Difference in temperature from 6°C to 0°C is 6°, and from 0°C to -7°C is 7°; difference between 6°C and -7°C is (6° + 7°) = 13°

11 **(D)** 10 per month = 120 per year; 291 for 3 years = 97 per year. 120 + 97 = 217

12 **(B)** If you divide 308 into 7 parts, each part would be twice as large as if you divided 308 into 14 parts.

13 **(C)** ¼ x \$30 = \$7.50, ⅓ x \$30 = \$10; spent \$17.50, left \$12.50.

14 **(C)** 20 centuries = 20 x 100 = 2 000 years, 11 decades = 11 x 10 = 110 years.

15 **(C)** 998 ÷ 27 = 36 + remainder →
The 998th line must therefore be on the 37th page.

```
      36
27 | 998
     81
     188
     162
      16
```

16 **(C)** Parallel lines are in the same plane and always the same distance apart; only 1 and 5 are parallel.

17 **(A)** Denominators are $1^3, 2^3, 3^3, 4^3, \ldots$, likely missing number is $5^3 = 125$

18 **(D)** Note 1km = 1 000m; 1ha = 100 x 100 = 10 000 m².
Area = 3 000 x 1 000m² = 3 000 000 ÷ 10 000 = 300ha.

19 **(C)** Distance travelled per week = (26.5 + 38.4 + 27 + 35) x 5 = 126.9 x 5 = 634.5km.

20 **(C)** Cost in instalments = 52 x \$15 = \$780
Cheaper by cash; saving = (\$780 - \$650) = \$130

21 **(C)** Number before multiplication by 4 should have been 624 ÷ 4 = 156;
after division by 4, answer is 156 ÷ 4 = 39

22 **(B)** ¾ corresponds to 96 marbles
i.e. ¼ corresponds to 96 ÷ 3 = 32 marbles
Thus $\frac{4}{4}$ = whole of Don's marbles = 32 x 4 = 128 marbles.

23 **(D)** 1cm³ = 1mL, and 1 000mL = 1L. Thus 1¼ L = 1 250mL = 1 250cm³

24 **(C)** All sides of an equilateral triangle are equal. Perimeter= 3 x 5½= 16½cm=16·5÷100m

25 **(B)** From bar graph, egg makes up 2½ parts out of 10; i.e. ¼ of sandwiches sold were egg; ¼ of 120 = 30.

26 (D) Obvious from sketches given below.

cone

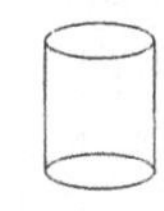
cylinder

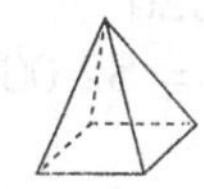
square pyramid

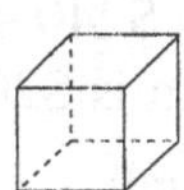
cube

27 (C) Sphere, cone, cube are solids (3 dimensional),
but circle is a plane figure (2 dimensional)

28 (A) Food and gas make up a quadrant, i.e. 25% of spent income.
Since food is 16%, gas must be 9%.

29 (C) 'OTHERS' make up a quadrant of the circle, i.e. 'OTHERS' make up 25% of spent income. 25% = $\frac{1}{4}$

30 (C) Rent, gas, clothing form half the circle; i.e. these make up 50% of spent income.

31 (A) The amount of money may change, but the percentage of that amount does not change.

32 (D)

	9%	of the total income spent is \$180
i.e.	1%	of this total income is \$180 ÷ 9 = \$20
Thus	16%	of this total income is 16 x \$20 = \$320.

33 (D)

$$\begin{array}{r} 1\,000\,000 \\ \underline{1\,000}\ - \\ \underline{999\,000} \end{array}$$

34 (D) The clock loses 4 minutes every hour, and thus after 1h (i.e. at 11a.m.) it will be correct. After 5 more hours (i.e. at 4p.m.) it will be 5 x 4 = 20 minutes slow.
Time on the clock then is 3:40p.m.

35 (D)

	8m	shadow cast by pole (or tree) 6m high
i.e.	1m	shadow cast by pole (or tree) $\frac{6}{8}$m high
Thus	3m	shadow cast by pole 3 x $\frac{6}{8}$ = $2\frac{1}{4}$m high

36 (B) 1 part cordial is mixed with 4 parts water.
8 parts cordial are mixed with 32 parts water.
Thus 8L cordial are mixed with 32L water to make 40L drink.

37 (D) Amount gold = 6% of 2 000 = $\frac{6}{100}$ x $\frac{2000}{1}$ = 120kg.

38 (C) Number of 10 minute intervals in 1hr = $\frac{60}{10}$
Number of cars produced per assembly line in 1hr = 8 x $\frac{60}{10}$
If 7 assembly lines, no. of cars = (8 x $\frac{60}{10}$) x 7

39 (B) Fraction = $\frac{1\frac{1}{2}}{24} = \frac{3}{48} = \frac{1}{16}$

40 (B) Jimmy has $\frac{4}{5}$ of \$80 = \$64

41 (D) Each hour the *big* hand moves through a complete rotation, i.e. 360°. From 1p.m. to 6p.m. on the same day, it rotates through 5 x 360° = 1800°.
(Note Each hour, the *small* hand moves through one twelth of 360° = 30°.
In 5h, it will rotate through 5 x 30° = 150°).

42 (A) QT is a diameter of the circle; its length is 2 x 6 = 12cm.
The sides of a regular hexagon are equal,
i.e. PQ = PU = UT = one sixth of 36cm = 6cm
Perimeter of shape 3 = 12 + 6 + 6 + 6 = 30cm

43 (A) 12% of a sum of money is \$240
i.e. 1% of this sum is \$240 ÷ 12 = \$20
Thus 100% i.e. whole sum is \$20 x 100 = \$2 000

44 (B) Each side of a square is equal;
thus FE = FB = BC = CE = 16m.

Each side of an equilateral triangle is equal;
thus AF = AB = BF = 16m, and
DE = DC = CE = 16m.

A B F 16m C E D

Perimeter of figure ABCDEF = 6 x 16 = 96m

45 (D) All angles of a square are 90° and of an equilateral triangle are 60°.
Angle x = 90° + 60° = 150°

PAPER 3

1 (A) $2^2 = 2 \times 2 = 4$, $2^3 = 2 \times 2 \times 2 = 8$; 4 + 8 =12

2 (D) Note 64.01 + 1.00 = 65.01; 64.01 - 1.00 = 63.01;
64.01 x 0 = 0; 64.01 ÷ 1 = 64.01

3 (C) Axes of symmetry shown on sketch

4 (D) **Method 1** Try each possible answer
For example if □ = 120, 4 x □ - 288 = 4 x 120 - 288 = 480 - 288 = 192
and thus □ ≠ 120, etc.

Method 2 Now 4 x □ - 288 = 52
Thus 4 x □ = 52 + 288 = 340
i.e. □ = 340 ÷ 4 = 85

5 (B) Braces ( ) are done before x÷, which are done in the order they occur.
Thus 6 x 4 ÷ 3 = 24 ÷ 3 = 8 and 6 x (4 ÷ 3) = 6 x $1\frac{1}{3}$= 8
Also 4 ÷ 5 x 2 = $\frac{4}{5}$ x 2 = $\frac{8}{5}$ and 4 ÷ (5 x 2) = 4 ÷ 10 = 2/5
Also 4 x 7 x 9 = 28 x 9 = 252 and (9 x 4) x 7 = 36 x 7 = 252
Also 27 ÷ 3 x 4 = 9 x 4 = 36 and (27 ÷ 3) x 4 = 9 x 4 = 36

6 (B) **Method 1** Noting 296 = 300 - 4
then 296 x 9 = (300 - 4) x 9 = 300 x 9 - 4 x 9 = 2 700 - 36

Method 2 By multiplication 296 x 9 = 2664
and by subtraction 2 700 - 36 = 2664

7 (D) Note 3 ÷ 10 = $\frac{3}{10}$; 13 x 10 = 130; now 1 ÷ $\frac{1}{10}$ = 10 and
thus 13 ÷$\frac{1}{10}$ = 13 x 10 = 130; 13 ÷ 10 = $\frac{13}{10}$ = $1\frac{3}{10}$

8 (C) I married at (61 - 42) = 19 years and graduated at (19 + 7) = 26 years.

9 (A) Time from 7:49a.m. to 8.00a.m. is 11 min. and then to 3:34p.m. is 7h34min.
Time away from home = 11min. +7h34min. = 7h45min.

10 (B) Tank holds $\frac{3}{4}$ x 2 400L = 1 800L
Amount used = $\frac{1}{4}$ x 1 800L = 450L, leaving 1 350L in tank.

11 (C) Temperature difference between -15°C and 0°C is 15°, and between 0°C and 15°C is 15°. Thus temperature difference between -15°C and 15°C is 15° + 15° = 30°.C

12 (B) $\frac{1}{10} = 0.1$, $\frac{3}{100} = 0.03$, $2 = 2.00$

$$\begin{array}{r} 0.1 \\ 0.03 \\ \underline{2.00} \\ \underline{2.13} \end{array}$$

13 (D) Note 20% of 10 = $\frac{1}{5}$ x 10 = 2; 2% of 100 = $\frac{1}{50}$ x 100 = 2; 40% of $\frac{1}{2}$ = $\frac{2}{5}$ x $\frac{1}{2}$ = $\frac{1}{5}$; 25% of 9 = $\frac{1}{4}$ x 9 = $2\frac{1}{4}$

14 (B) The answer can easily be seen from sketches.

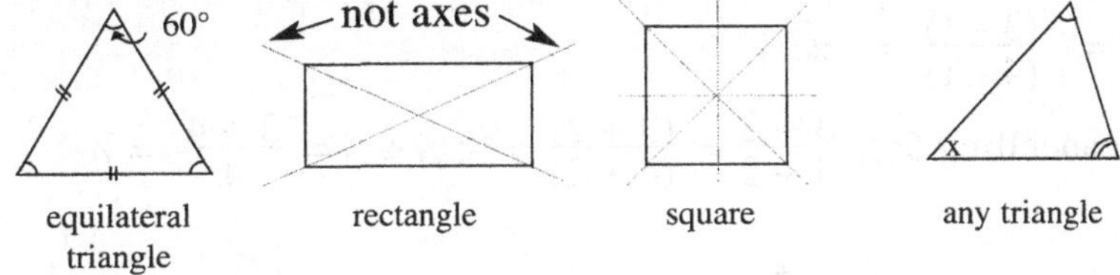

15 (C) Fraction planted = $\frac{1}{2} + \frac{1}{3} = \frac{3}{6} + \frac{2}{6} = \frac{5}{6}$, which leaves $\frac{1}{6}$ not planted.

16 (C) A speed of 20 km in $\frac{1}{4}$h means 80km in 1h.

17 (D) Now 25% = $\frac{1}{4}$ and thus 12.5% = $\frac{1}{2}$ of $\frac{1}{4}$ = $\frac{1}{8}$.

Method 1 Try the possible answers given, taking the simplest namely $800 first. If price before discount is $800, then price after discount is $800 - $\frac{1}{8}$ of $800, i.e. $800 -$100 = $700.

Method 2

Discounted price = original price - discount
= original price - $\frac{1}{8}$ original price
= $\frac{7}{8}$ original price

That is, original price = $\frac{8}{7}$ of discounted price
= $\frac{8}{7}$ x $700
= $800

18 (C) Note $3\frac{1}{2}$ minutes = (3 x 60 + 30) = 210 seconds.
In 10 seconds the girl types 3 words
In 210 seconds, (i.e. 10 x 21) she types 3 x 21 = 63 words

19 (B) We cut the figure into rectangles as shown.

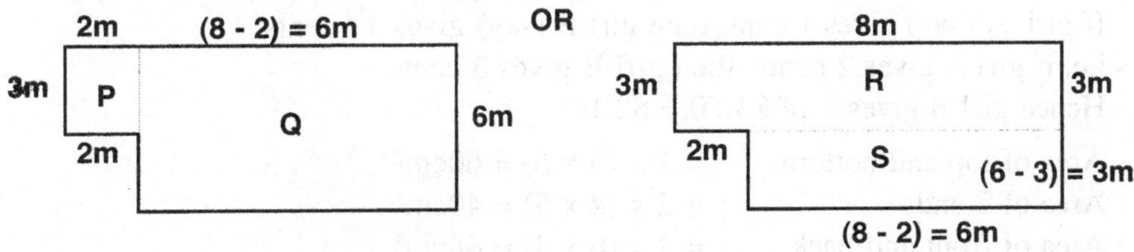

Method 1 Area of P = 3 x 2 = 6m² and of Q = 6 x 6 = 36m²
Method 2 Area of R = 3 x 8 = 24m² and of S = 3 x 6 = 18m²

20 (D) Error = 16.4m x 25 - 16.4m x 5
= 16.4m x (25 - 5), i.e. 16.4m x 20
= 328m

21 (B) Discount = $\frac{15}{100}$ x \$210 = \$$\frac{63}{2}$ = \$31.50
Sale Price = \$210 - \$31.50 = \$178.50

22 (A) $2\frac{3}{4} - \frac{1}{4} = 2\frac{1}{2}$; $1\frac{1}{2} + 2\frac{1}{2} = 4$

23 (B) Work out cost per litre for each case.
500mL for \$1 means 1L costs \$1 x 2 = \$2
5L for \$9.90 means 1L costs \$9.90 ÷ 5 = \$1.98
3L for \$6.03 means 1L costs \$6.03 ÷ 3 = \$2.01

24 (B) Note $\frac{3-1}{4-1} = \frac{(3-1)}{(4-1)} = \frac{2}{3}$;
$\frac{3 \times 5}{4 \times 5} = \frac{3}{4}$ (cancelling 5s); $\frac{3+2}{4+2} = \frac{(3+2)}{(4+2)} = \frac{5}{6}$; $\frac{3}{4} + \frac{4}{4} = \frac{3+4}{4} = \frac{7}{4}$

25 (C)

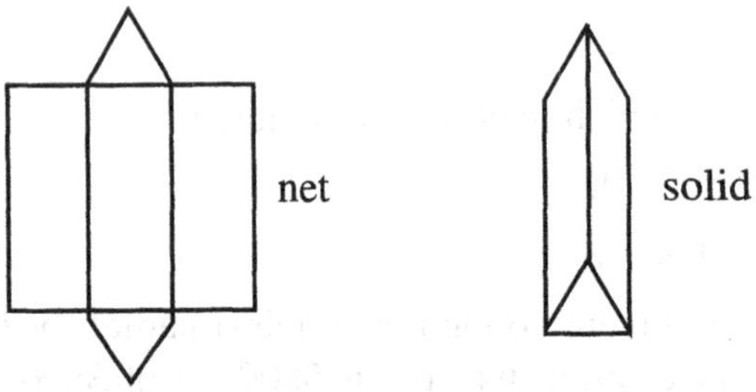

net   solid

The net produces the solid shown, which is a triangular prism.

26 (C) Only the pair 20, 5 give a sum of 25 and a product of 100.

27 (A) His weekly income consists altogether of (3 + 4 + 1) = 8 parts, of which 4 parts are spent on food.
Amount spent on food = $\frac{4}{8}$ of \$600 = \$300.

28 (D) In each given pattern, the dots move together 2 spaces in an anti-clockwise direction.

29 (B) The length of AB is the sum of the diameters of the 3 circles, i.e. (4 + 9 + 14) = 27cm.

30 (D) Area given triangle
= ½ base x perpendicular height
= ½ x 6 x 6, i.e. 18cm^2
OR
area triangle = ½ area of square shown
= ½ x 6^2, i.e. 18cm^2

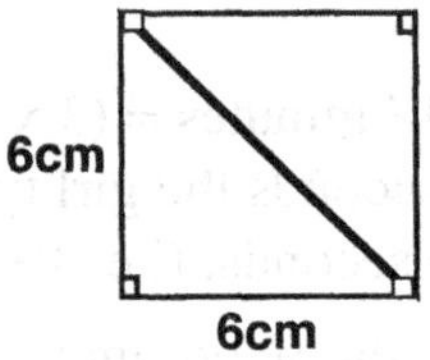

6cm
6cm

31 (B) If girl A (say) gives 1 cent, then girl B (say) gives 1½ cents
i.e. if girl A gives 2 cents, then girl B gives 3 cents
Hence girl B gives $\frac{3}{5}$ of \$3.50 = \$2.10

32 (A) Area of top and bottom = 2 x (5 x 6) = 60cm^2
Area of 2 ends = 2 x (4 x 5) = 40cm^2
Area of front and back = 2 x (6 x 4) = 48cm^2
Total surface area = (60 + 40 + 48) = 148cm^2

33 (D) Total cost for 3 boys = (\$6 + \$10.50 x 3 + \$2 x 3 + \$6 x 3) = \$61.50
Each boy's share = \$61.50 ÷ 3 = \$20.50

34 (C) Area ΔPQR = area ΔTRS = ½ x 3 x 4 = 6 cm^2. Total area = 12 (cm^2)

35 (D) The triangles PQR, TSR have the same dimensions and if cut out would obviously fit on one another.
They are obviously right-angled; the sides PR, TR are equal; the angles PRQ, TRS are equal.
However, since $\angle TRS = \angle PRQ$, then
$\angle QPR + \angle TRS = \angle QPR + \angle PRQ = 90°$, in ΔPQR
i.e. $\angle QPR + \angle TRS \neq 180°$

36 (D) This is easiest done by making a table as below, and filling it in, starting with Jane.

Paul	Jane	Michelle	George
	14		

→

Paul	Jane	Michelle	George
7	14	2	2

Total no. of pens = 7 + 14 + 2 + 2 = 25

37 (C) Paul has 7 pens out of the 25 pens. Percentage = $\frac{7}{25} = \frac{28}{100}$ = 28%

38 (A)

Area of side walls	= 2 x (4 x 3)	= $24m^2$
Area of front & back walls	= 2 x (5 x 3)	= $30m^2$
Area of ceiling	= 5 x 4	= $20m^2$
Area of windows and door	= (2 x 2.5 +2)	= $7m^2$
Area to be painted	= 24 + 30 + 20 - 7	= $67m^2$

39 (B) 60 minutes cost 36c
1 minute costs $\frac{36}{60}$c
4 minutes cost $\frac{36}{60}$ x 4c = $\frac{12}{5}$c = 2.4c

40 (A) Sum = 0.81 + 0.78 = 1.59

41 (C) Note 80 cm = 0.8m
and 50 cm = 0.5m
Volume of piece shaded is given by

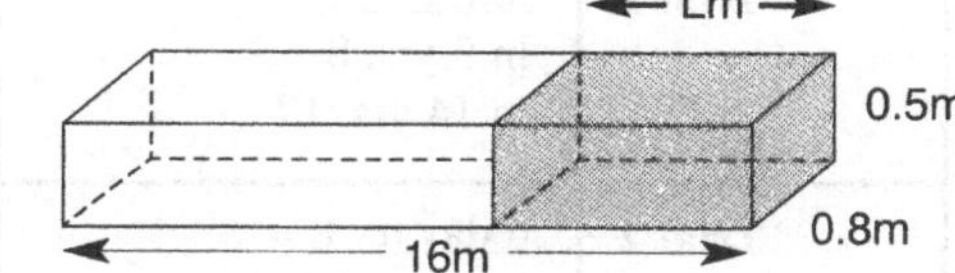

V = L x B x H
i.e. 2 = L x 0.8 x 0.5
i.e. 2 = L x 0.4, noting 0.8 x 0.5 = 0.40

Thus L = $\frac{2}{0.4} = \frac{2 \times 10}{0.4 \times 10} = \frac{20}{4} = 5$

i.e. length of piece is 5m, which leaves a length of (16 - 5) = 11m of beam.

42 (B) Note 11 = 11.00

$$\begin{array}{r} 11.00 \\ 0.01\ - \\ \hline 10.99 \end{array}$$

43 (A) 75% of 75 = $\frac{3}{4}$ x 75 = $\frac{225}{4}$ = $56\frac{1}{4}$

44 (A) Perimeter of regular hexagon is 6 x 5 = 30cm
Thus perimeter of square is 30cm
Side of square = 30 ÷ 4 = $7\frac{1}{2}$cm

45 (C) Area of larger square = $4^2 = 16cm^2$
Area of each small square = $(0.2)^2 = 0.04cm^2$

No. of small squares = $\frac{16}{0.04} = \frac{16 \times 100}{0.04 \times 100} = \frac{1600}{4} = 400$

PAPER 4

1 (C) Note 6 x 7 = 42; thus 0.6 x 7 = 4.2 (1 decimal place)

2 (D) Note $\frac{2}{5} = \frac{4}{10}$ = 0.4, $\frac{1}{4}$ = 0.25

3 (C) We do multiplication before addition. 14 x 2 + 2 x 15 = 28 + 30 = 58

4 (D) Note 4 892 [-] 3 127 = 1 765, 1 765 [+] 3 127 = 4 892

5 (B) One half of 80 = 40; by trial one fifth of 200 = 40

6 (C) Note 6.209 ≠ 6.29; 20.7 ≠ 2.7; 7.890 = 7.89; 790 ≠ 79

7 (D) Note 3h45min. = 25min. + 3h20min.
11:35a.m. + 25min. = 12:00 noon
noon + 3h20min. = 3:20p.m.

8 (B) Order is RYPB, RYPB, RYPB, … in sets of 4 colours.
Now 58 ÷ 4 = 14 rem 2; i.e. 14 sets of 4 colours + 2 left over; i.e. 58th marble is Y.

9 (D) A number is divisible by 9 if the sum of the digits is divisible by 9.
The digits in 12 □ 7 add to 10 + □ , and hence □ = 8.

10 (B) Try each possible answer to see if 16 x □ = 9 x 5 + □
Thus, if □ = 5, then 16 x □ = 16 x 5 = 80 and 9 x 5 + □ = 45 + 5 = 50, and so on.

11 (A) We use trial and error, and draw up a table as shown, trying various ages for my son.

	Son's age in years	My age in years	Sum of ages
Trial 1	today try 3 in 5 yrs, 8 in 14 yrs, 17	- 3 x 8 = 24 24 + 9 = 33	- - 17 + 33 = 50 - too small
Trial 2	today try 8 in 5 yrs, 13 in 14 yrs, 22	- 3 x 13 = 39 39 + 9 = 48	- - 22 + 48 = 70 - too large
Trial 3	today try 6 in 5 yrs, 11 in 14 yrs, 20	- 3 x 11 = 33 33 + 9 = 42	- - 20 + 42 = 62 - correct

Son's age today 6 years, my age today (33 -5) = 28 years; sum of ages 34.

12 (C) In 5 minutes, car travels 6km
In 60 minutes, car travels 6 x 12 = 72km

13 (C) Perimeter of triangle = (7 + 14 + 15) = 36cm
Perimeter of equilateral triangle = 36cm
Side of equilateral triangle = 36 ÷ 3 = 12cm

14 (B) Number in descending order 9542 and in ascending order 2459.
Difference = 9542 - 2459 = 7083

15 (C) Quotient is 0.05

$$2\overline{)0.10}$$
$$0.05$$

16 (D) (11 - 11) = 0; $\frac{11 \times 11}{11 \div 11} = \frac{121}{1}$ = 121; (11 + 11) = 22. Result = 0 + 121 - 22 = 99

17 (A) Try each possible answer
Thus, if the number is 7, the sum of it and 5 is 12; then 6 more than twice this sum is 6 + 24 = 30.
For practice, try 10½, 9, 11 similarly.

18 (D) 1 cm represents 25km
7¼cm represents 25 x 7¼ = 25 x 7 + 25 x ¼ = 175 + 6¼ = 181.25km.

19 (C) Direction is now NE.

Before: N, E, S, W, A.C.
After: N, NE, E, S, W, 135°

20 (C) 1km = 1 000m; 0.5km = 500m; 0.2km = 200m
Area = 500 x 200m² = 100 000m²
= 100 000 ÷ 10 000ha = 10ha (note10 000m² =1ha)

21 (C) No. of trees = 1 + 1000/20 = 1 + 50 = 51; don't forget the tree at the start

22 (D) 1m = 100cm; 7m = 700cm, 8m = 800cm, 10m = 1000cm
V = L x B x H = 800 x 1000 x 700cm³ = 560 000 000cm³,
note there are (2 + 3 + 2) = 7 zeros

23 (D) Third number = 2688 ÷ (14 x16) = 2688÷ 224 = 12.

24 (D) Number passed = (650 - 130) = 420
Percentage passed = 420/650 x 100 = 80%, on cancelling.

25 (A) Area surface = 700 x 700cm²; area each tile = 20 x 25cm²

No. of tiles = $\frac{700 \times 700}{20 \times 25}$ = 35 x 28 = 980

26 (C) 3% of 100 = 3/100 x 100 = 3; 100% of 3 = 100/100 x 3 = 3

27 (D) One revolution of wheel = circumference of wheel = 1.61m
Distance travelled = 25 x 1.61m = 40.25m, on multiplying

```
 1.61
   25 x
  805
 322
40.25
```

28 (C) Number = $\frac{10\ 000\ 000}{100}$ = 100 000 (NB 7 - 2 = 5 zeros)

29 (B) Difference in temperature = 24.9°C - 9.6°C = 15.3°C

Average hourly increase = $\frac{15.3°C}{6}$ = 2.55°C

```
6 ) 15.30
     2.55
```

30 (B) 7/10 = 0.7

31 (C) ¾ x 48 000L = 36 000L. Time = $\frac{36\ 000}{900}$ = 40min.

32 (D) Note start at one corner and count around figure to reach corner again.

33 (B) Area of each square = 1m²

34 (A) Cost apples $2.40, meat $7.35, rolls $2.40, milk $2.25. Total $14.40

35 (C) Mark for 6 subjects = 6 x 72 = 432, and for 7 subjects is (432 +100) = 532.
Average for 7 subjects is 532/7 = 76

36 (A) Note 2km = 2 000m = 200 000cm
No. of steps for Mary = $\frac{200\ 000}{40}$ = 5 000
No. of steps for Jim = $\frac{200\ 000}{50}$ = 4 000

37 (C) Royalty = 20% of (160 000 x $13.80) = ⅕ x 1 600 x $1380
= 320 x $1380 = $441 600, on multiplying.

38 (A) If ⅔ below ground, then ⅓ is above ground.
Thus ⅓ pole = 15m and whole pole = 3 x 15 = 45m

39 (C) Try each possible answer.
Thus if 11h, she earned 11 x $1.60 = $17.60 and so on.

40 (A) 40% of number is 64
i.e. 10% of number is 64 ÷ 4 = 16
and 70% of number is 16 x 7 = 112

41 (C) From graph, 2kg lead costs $10 and 2 kg aluminium costs $15; total is $25.

42 (B) $15 buys 2kg aluminium
i.e. $15 000 buys 2 x 1 000 = 2 000kg aluminium

43 (C) 10 000kg aluminium at $7.50 per kg costs $7.50 x 10 000 = $75 000, and this leaves $25 000 to buy lead at $5 per kg.
Amount purchased = $\frac{25\ 000}{5}$ = 5 000kg.

44 (C) Change all units to cm first.
Vol. of container = L x B x H = 800 x 500 x 300 cm^3
Vol. of each box = 200 x 50 x 20 cm^3
No. of boxes = $\frac{800 \times 500 \times 300}{200 \times 50 \times 20}$ = 4 x 10 x 15 = 600

45 (C) **Method 1** Sums in pairs are 12, i.e. 1 + 11 = 12,
3 + 9 = 12, 5 + 7 = 12, 7 + □ = 12 and thus □ = 5
Method 2 Odd terms form pattern 1, 3, 5, 7
Even terms form pattern 11, 9, 7, □
Thus □ = 5

PAPER 5

1 (D) Note 7 = 7.00 ⟶ 7.00 − 3.48 = 3.52

2 (D) 437 x 6 = 2622; 4.37 x 6 = 26.22.

3 (D) Best to change units to metres. Noting 100cm = 1m, then 250cm = 2½m, 50cm = ½m.
P = 2(L + B) = 2(2½ + ½) = 2 x 3 = 6m

4 (C) A = L x B = 2½ x ½ = 5/2 x ½ = 5/4 = 1¼ = 1.25m^2

5 (D) 1 million cents = $1 000 000 ÷ 100 = $10 000 ⟶ 1 000 000 − 10 000 = 990 000
Difference =$1 000 000 - $10 000 = $990 000

6 (B) Answer = $\sqrt{6^2 + 8^2} = \sqrt{36 + 64} = \sqrt{100}$ = 10

7 (B) For cube, $V = s^3$ and $s = \sqrt[3]{V}$ (s = side)
Side of cube, of volume $27cm^3$, is $\sqrt[3]{27} = 3cm$ (since $3^3 = 27$)
Area of each square face = $3^2 = 9cm^2$
Surface area of cube = 6 x 9 = $54cm^2$

8 (A) Note division carefully! $4\overline{)0.32}$ = 0.08

9 (A) Length of circumference = $3\frac{1}{7}$ x diameter of wheel
i.e. length = $3\frac{1}{7}$ x 70 = 3 x 70 + $\frac{1}{7}$ x 70 = 210 + 10 = 220cm

10 (C) Each revolution of the wheel in cycling carries the cycle forward 220cm.
Now 110m = 11 000cm and no. of revolutions is 11 000 ÷ 220 = 50

11 (D) Sum of four numbers is 4 x 4.5 = 18, but sum is (2 + □ + 7 + 6) = 15 + □. Thus □ = 3.

12 (C)
8kg sugar is needed for 10kg fruit
1kg sugar is needed for $\frac{10}{8}$kg fruit
i.e. 6kg sugar is needed for $\frac{10}{8}$ x 6 = $7\frac{1}{2}$kg fruit

13 (B) 3/100 + 6/10 = 3/100 + 60/100 = 63/100 = 63%

14 (B) Area of 2 side walls = 2 x (5 x 3) = $30m^2$
Area front & back walls = 2 x (7 x 3) = $42m^2$
Total area 4 walls = $72m^2$

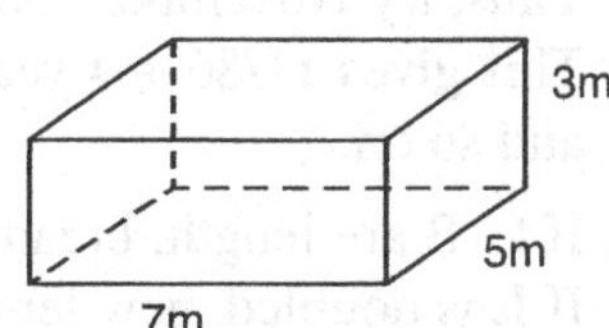

15 (C) V = L x B x H = 7 x 5 x 3 = $105m^3$
Capacity of room = 105 x 1 000 = 105 000L

16 (B)
If $2\frac{1}{2}$ litres juice cost \$1.80
then 5 litres juice cost \$3.60
i.e. 1 litre juice costs \$3.60 ÷ 5 = 72c
and$1\frac{1}{2}$ litres juice cost (72c + 36c) = \$1.08

17 (C) Note $\frac{75.32}{12.62} \div \frac{78}{13} = 6$
Result given $\div\ 6 \times \frac{140}{80} = \frac{84}{8} \div 10$

18 (C) Amount vaccine used = 4 586 x $\frac{1}{2}$mL = 2 293mL = 2.293L

19 (C) Add 4 000, 700, 9

20 (C) Total for the 5 measurements = 5 x 13 = 65
Total for the 7 measurements = 7 x 10 = 70
Average for the 12 measurements = (65 + 70) ÷ 12 = 11.25

21 (D) If a number is divisible by 3 (i.e. a multiple of 3) and is divisible by 4, then it is divisible by 3 x 4 = 12.
Numbers less than 40 divisible by 12 are 12, 24, 36.

22 (C) By data, if a, b are numbers then aPb =4 x (a + b)
Thus 5P6 = 4 x (5 + 6) = 4 x 11 =44

23 (C) Third angle is 360° -(107° + 208°) = 45° ⟶
Third sector = $\frac{45}{360}$ x 100% = $12\frac{1}{2}$% of the pie chart

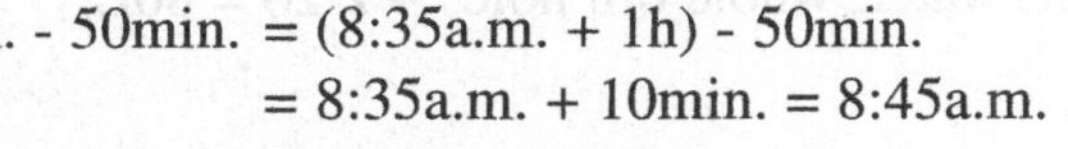

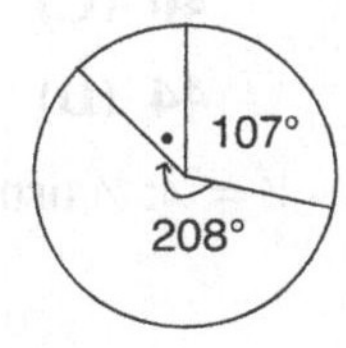

24 (A) 9:35a.m. - 50min. = (8:35a.m. + 1h) - 50min.
= 8:35a.m. + 10min. = 8:45a.m.

25 (B) Multiplication is done before addition.
Note 0.1 x 40 = $\frac{1}{10}$ x 40 = 4. Answer = 28 + 4 = 32

26 (B) Number is $\frac{1}{2}(0.4 + 1.68) = \frac{1}{2}(2.08) = 1.04$

27 (C) Easy to see if you draw sketches and draw lines of symmetry (if any).

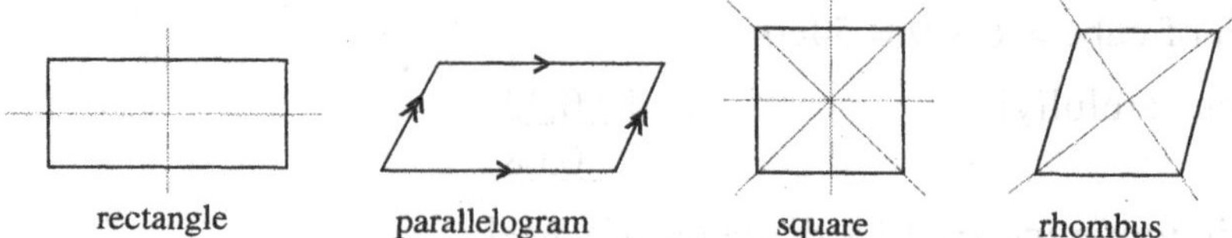

Note a square has 4 axes of symmetry, a rectangle and a rhombus 2 each, a parallelogram has none.

28 (C) Direction from P to Q is NE, and from Q to P is SW.

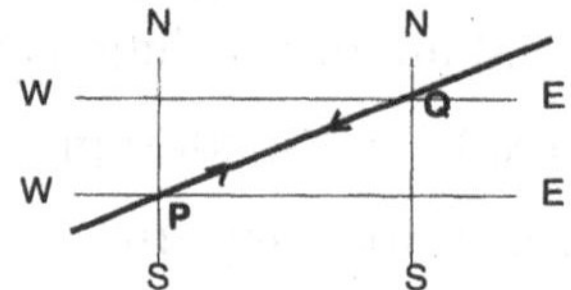

29 (D) Fraction = $\frac{300}{60\,000} = \frac{3}{600} = \frac{1}{200}$

30 (C) Try each possible answer.
Thus, try November 1986 + 4 years 5 months
This gives 11/86 + 4 years + 5 months = 11/90 + 5monthsi.e. 4/91,
and so on.

31 (D) If L, B are length, breadth of original rectangle, then this has area A = L x B.
If L is doubled, new length is 2L and if B is tripled, new breadth is 3B.
Area of new rectangle = 2L x 3B = 6 x L x B; thus the original area is multiplied by 6.

32 (B) The bottle and 16 pills have mass 140g
The bottle and 10 pills have mass 125g
i.e. (16 - 10) = 6 pills have mass (140 - 125) = 15g
Thus 2 pills have mass 15 ÷ 3 = 5g
and hence16 pills have mass 8 x 5 = 40g
Since the bottle and 16 pills have mass 140g, then the bottle has mass (140 - 40) = 100g.

33 (D) 20% of \$8 = $\frac{1}{5}$ x \$8 = \$1$\frac{3}{5}$ = \$1.60

34 (B) $\frac{3}{4} = 0.75$, 74% = $\frac{74}{100} = 0.74$

35 (D) Even divisors of 32 are 2, 4, 8, 16, 32; their sum is (2 + 4 + 8 + 16 + 32) = 62

36 (A) There are 100 numbers in the set; if we pair them thus
(100 - 99) + (98 - 97) + ... + (4 - 3) + (2 -1),
then there are 50 pairs each of value 1. Total value 50 x 1 = 50

37 (D) We can find a, then b and then x →

		12
9	a	13
b	x	8

38 (D) Try each possible answer.
If Jane gives \$5 to Neil and \$1 to Mary,
we obtain line 2 - not correct, →
since Neil does not have \$5 more than Jane and so on.

N	J	M
\$30	\$30	\$30
\$35	\$24	\$31

39 (C) **40 (C)** **41 (B)** Dots move 2 sectors anti-clockwise around the circle.

43 (D) **44 (D)**

45 (D) $\frac{3}{4} - \frac{1}{2} = \frac{1}{4}$; $\frac{1}{4}$ urn holds 20L water; whole urn holds 4 x 20 = 80L.

Solutions not given to Questions similar to previous ones.

PAPER 6

1 (A) **2 (B)** 3.97 x 2.134 ÷ 4 x 2 = 8

3 (D) Change to same units, here mm.
7mm on map represents 350m = 35 000mm distance
1mm on map represents 350 000 ÷ 7 = 50 000mm distance
Scale is 1:50 000

4 (D) In 60 min. car travels 78km
In 20 min. car travels 78 ÷ 3 = 26km

5 (D) Shaded area = 12 x 8.5 - 6 x 4.5 = 102 -27 = 75m^2

6 (B)

7 (C) Count numbers possible thus: ⟶ (5) ( ) ( ) ( )
5137, 5173, 5317, 5371, 5713, 5731

8 (A)

9 (D) Rainfall for 12 months = 12 x 9 = 108 cm
Rainfall for December = (108 - 93) = 15cm

10 (A) **11 (C)** **12 (B)**

13 (D) Charge for 50km = 50 x 90c = $45
Total cost of bus for journey = $135 + $45 = $180
Cost per passenger = $180 ÷ 40 = $4.50

14 (C) Zinc is 7 parts out of (7 + 13) = 20 parts
Noting 1t = 1 000kg, amount zinc = $\frac{7}{20}$ x 1 000 = 350kg

15 (D) In 632, value of 6 is 600, 3 is 30 and 2 is 2.

16 (A) **17 (B)**

18 (A) If 90% of a number is 360
then 10% of the number is 360 ÷ 9 = 40
and the whole number (i.e. 100%) = 40 x 10 = 400

19 (D) 7c buys 2 sheets
1c buys $\frac{2}{7}$ sheets
$1.54 =154c buys $\frac{2}{7}$ x 154 = 44 sheets

20 (A) Gain is 6c. Gain as fraction of S.P. = $\frac{6}{30}$ = $\frac{1}{5}$

21 (B) Try each possible answer
<u>Thus 1 day</u> = 24h = 24 x 60 min. = 24 x 60 x 60s = 86 400s
At $1 per s, amount spent = $86 400. not correct
<u>If 10 days</u>, spent $864 000, which is nearly $1M.
Obviously 10 weeks = 70 days and 1 year = 365 days are too big.

22 (D) Noting 1m = 100cm, then
Vol. of water = 300 x 400 x 500cm^3 = 300 x 400 x 500mL
= $\frac{300 \times 400 \times 500}{1000}$ L = 30 x 40 x 50L
Weight of water = 30 x 40 x 50kg = $\frac{30 \times 40 \times 50}{1000}$ t = 60t

23 (A) Noting 1km = 1 000m, 1ha = 10 000m² then
area farm = 2 000 x 1 600m² = $\frac{2\,000 \times 1\,600}{10\,000}$ ha = 320ha

24 (B) **25 (B)** **26 (C)**

27 (B)
5.5 km uses 500mL = ½L petrol
11 km uses 2 x ½ = 1L petrol
1 km uses $\frac{1}{11}$L petrol
218 km uses $\frac{1}{11}$ x 218 = 19$\frac{9}{11}$L ∻ 20L

28 (D) Best to sketch each figure

29 (D) No. of cups from cask = $\frac{3000}{160}$ = $\frac{150}{8}$ = 18¾

30 (B) Try each possible answer. ⟶
Thus if 18 squares, get line 1 and so on.

triangles	squares	sum
12	18	30

31 (A) Area large square GDEF = 9² = 81cm²
Area of half-square ABCH = 89 - 81 = 8cm²
Area smaller square = 2 x 8 = 16cm²
Side of smaller square = $\sqrt{16}$ = 4cm

32 (D) Numbers with digit 7 from 1 to 120 are
7, 17, 27, 37, 47, 57, 67, 70, 71, 72, 73, 74, 75, 76, 77, 78, 79, 87, 97, 107, 117
: a total of 22 digits 7 (note 2 digits in 77)

33 (A) At 2:30, minute hand is on 6
and hour hand is midway between 2 and 3.
Now each hour division represents 360° ÷ 12 = 30°
By counting, required angle= 3½ x 30° = 90° + 15° = 105°

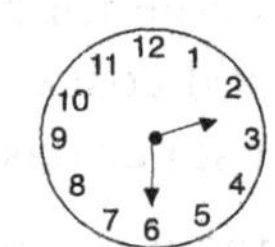

34 (D) Required number is 1 more than lowest number divisible by 6, 8, 9, 24
(i.e. L.C.M. of 6, 8, 9, 24)
Now 6 = 2 x 3, 8 = 2³, 9 = 3², 24 = 2³ x 3. L.C.M. = 2³ x 3² = 72
Required number is 72 + 1 = 73

35 (B) Area square = area rectangle = 16 x 9 = 144cm²
Side square = $\sqrt{144}$ = 12cm

36 (B) **37 (D)** cancel **38 (A)** **39 (A)** **40 (B)**

41 (D) 3 children's fare = 3 x $1.80 = $5.40
2 adult's fare = $16 - $5.40 = $10.60

42 (D) Try each possible amount.
Thus if 35c, change is $2 - 35c = $1.65, which needs at least 4 coins.

43 (D) 51, 57 are both divisible by 3; 49 by 7

44 (C)

45 (C) Try each possible answer.
Thus if U = 8, V = 6, W = 5 and so on.

```
  283
   34 x
 1132
 8490
 9622
```

PAPER 7

1 (D)

2 (A) First girl gets $2 out of each ($2 + $3 + $6) = $11. Her share $\frac{2}{11}$ x $165 = $30

3 (A)

4 (C) On map 1mm represents 20km
i.e. 1cm represents 20 x 10 = 200km
and 2.25cm represents 200 x 2.25 = 2 x 225 = 450km

5 (B) 20km distance represented by 1mm
i.e. 1km distance represented by $\frac{1}{20}$mm
Thus 700km represented by $\frac{1}{20}$ x 700 = 35mm = 3.5cm

6 (B) 0.7 of $23 = 7/10 x $23 = $161/10 = $16.10

7 (A) All dimensions shown can be obtained from given figure.
To find perimeter, start at a corner and go around the sketch.

8 (B) Divide figure into rectangle as shown.
Area figure = 3 x 8 + 4 x 6 + 3 x 8 = 24 + 24 +24 =72cm^2

9 (C) In circle OA is a radius, AC is a diameter,
AB is a chord, APB is an arc.

10 (C) In circle, area BDE is a segment, AFC is a semicircle,
AOB is a sector, COD is a quadrant.

11 (B) A triangle with no sides equal (or no angles equal) is called a scalene triangle, with 2 sides equal is an isosceles triangle and with 3 sides equal is an equilateral triangle.

12 (A) A 73km trip costs $14.60 = 1460c
i.e. a 1km trip costs $\frac{1460}{73}$ c =20c
Thus, a 65km trip costs 20c x 65 = $13

13 (A) Draw a sketch.
Area = 2 x (15 x 10) + 2 x (12 x 10) + 15 x 12 = 300 + 240 + 180 = 720m^2

14 (A) Vol. of air = 15 x 12 x 10 = 1 800m^3

15 (B)

16 (A) Noting 90 min = 1½h, then no. of revolutions = $\frac{24}{1\frac{1}{2}} = \frac{48}{3} = 16$

17 (C) Sequence can be written as $\frac{1}{2}, \frac{4}{4}, \frac{9}{8}, \frac{16}{16}, \frac{25}{32}, \frac{36}{64}$
Numerators are $1^2, 2^2, 3^2, 4^2, 5^2, 6^2$, and denominators are $2^1, 2^2, 2^3, 2^4, 2^5, 2^6$.
The next term is probably $7^2/2^7 = \frac{49}{128}$.

18 (D) Try each possible answer.

19 (D) Note 57 x 86 = 57 x (80 + 6) = 57 x 80 + 57 x 6

20 (C) No. of pieces = $\frac{300}{34} = \frac{150}{17} = 8\frac{14}{17}$. No. of complete pieces = 8

21 (C) 80% = $\frac{4}{5}$; no. of students = 3

22 (C) No. of students with 3, 4 or 5 marks is (5 + 3 + 1), i.e. 9

23 (D) No. of students with 2, 3 or 4 marks is (6 + 5 + 3) i.e. 14, out of total no. of students, i.e. (2 + 3 + 6 + 5 + 3 + 1) = 20.
Percentage is $\frac{14}{20}$ x 100 = 70%

24 (C) Use trial and error and form a table thus, taking various values for the breadth.
Try breadth 4m, obtain line 1; not correct.
Try 10m, and so on.

length	breadth	area	perimeter
1m	4m	$4m^2$ - too small	10m
7m	10m	$70m^2$ - too large	34m
5m	8m	$40m^2$	26m

25 (D) 8 x 8 x 8 = 512 = 500 to nearest hundred;
(note 510 is to nearest ten not to nearest hundred)

26 (D) Since in every 4 years, there is a leap year, then over a long period, it is better to take 1 year, not as 365 days, but as 365¼ days.
No. of days he lived = 80 x 365¼ = 80 x 365 + 80 x ¼ = 29 200 + 20
= 29 220 ≑ 29 000

27 (B) Area each floor = 25 x 16 = $400m^2$
No. of floors = 5 600 ÷ 400 = 14

28 (A) Enough food for 14 pigs for 16 days
i.e. enough food for 1 pig for 16 x 14 days
Thus enough food for 8 pigs for $\frac{16 \times 14}{8}$ = 28 days

29 (D) Total marks for 12 girls = 12 x 64 = 768
Total marks for whole class = 20 x 60 = 1 200
Total marks for 8 boys = 1 200 -768 = 432
Average mark for 8 boys = 432 ÷ 8 = 54

30 (C) The net can be folded to form a cube of side 10cm.
There are 12 edges and length of these edges is 12 x 10 =120cm

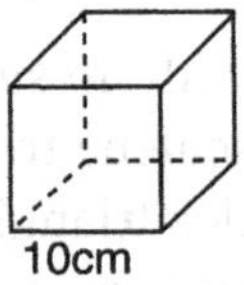

10cm

31 (A) By trial with whole numbers,
we can obtain the side dimensions shown.
Perimeter ABCD = 2 (11 +14) = 50cm

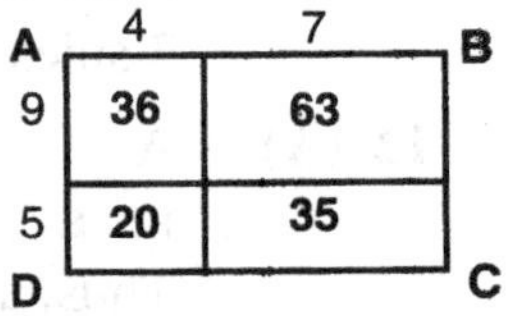

	4	7
9	**36**	**63**
5	**20**	**35**

(A top left, B top right, C bottom right, D bottom left)

32 (B) No. of spokes = 360/24 = 15

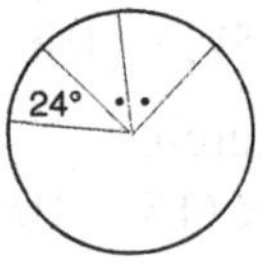

24°

33 (C) Noting 1m = 1 000mm, then
Area sheet = 1 000 x 1 000 = 1 000 $000mm^2$
Area of each small square = 1 x 1 = $1mm^2$
No. of squares = 1 000 000 ÷ 1 = 1 000 000
Each square has length 1mm, so distance along a line formed by 1 000 000 squares is 1 000 000mm = 1 000m

34 (A) **35 (D)** Sum = 12 + 23 + 34 + 45 + 56 + 67 + 78 + 89 = 404

36 (A) Value = (3) x (17) + (12) x (20) = 51 + 240 = 291

37 (C) Each small square is $\frac{1}{5}$ sq. unit.

38 (C) **39 (D)**

40 (D) Copy each number given on a sheet of paper, and rotate (on the desk) through a half turn i.e. 180°.

41 (B) 100mm = $\frac{1}{10}$m. Vol. = 6 x 4 x $\frac{1}{10}$ = $\frac{24}{10}$ = $2.4m^3$

42 (A) Value ÷ 5 x 0.8 = 4.0 = 4

43 (C)

44 (B) There are 20 whole numbers from 2 to 21, of which there are 7 multiples of 3 (namely 3, 6, 9, 12, 15, 18, 21). Percentage = $\frac{7}{20}$ x 100 = 35%

45 (C) Try each possible answer.
Thus, if dimensions (i.e. length & breadth) are 4cm, 7cm respectively, then perimeter is 22cm and area is 28cm^2; not correct.

PAPER 8

1 (C) In 474, the 4's represent 400, 4; difference 396.

2 (D) Value ÷ 60 x $\frac{60}{6}$ x $\frac{0.4}{0.4}$ = 60 x 10 = 600

3 (D) Percentage = $\frac{29}{40}$ x 100 = $\frac{145}{2}$ = 72½%

4 (B) Time = (2½ x 40 + 30)min = 130min = 2h10min

5 (D) Taking first number from top and second number from side, them to find say 5Ø2 we find the number where the column headed 5 intersects the row starting 2; the number there is 4, i.e. 5Ø2 = 4

6 (A) (5Ø2)Ø3 = 4Ø3, (using Q5) = 0

7 (D) Trying □ = 1, 2, 3, 4, 5 we see none are such that □Ø3 = 2

8 (D) Trying □ = 1, 2, 3, 4, 5 we see that if □ = 1 or 5, then 1Ø1 = 1 and 5Ø5 = 1

9 (A) No. of axes of symmetry for a regular hexagon is 6 and for an isosceles triangle is 1.

10 (A) Area = 7.5ha = 7.5 x 10 000m^2 = 75 000m^2
Other side = Area/length = 75 000 ÷ 600 = 125m

11 (D) An acute angle is less than 90°; an obtuse angle is between 90° and 180°; a reflex angle is between 180° and 360°.

12 (D) Be careful here! A quadrilateral with 4 sides equal only is a rhombus; with 4 right angles only is a rectangle; with opposite sides parallel is a parallelogram, with 4 equal sides and 4 right angles is a square.

13 (C) 200 sheets measure 48mm thick
1 sheet measures $\frac{48}{200}$ = $\frac{24}{100}$ = 0.24mm thick

14 (C) Vol. of cube = Vol. of prism = 9 x 8 x 3 = 216cm^3
Side of cube = $\sqrt[3]{216}$ = 6cm, noting 6^3 = 6 x 6 x 6 = 216

15 (A) 3 x 2 x 4 = 24; 3 x 0.02 x 4 = 0.24

16 (B) 357 619 lies between 357 000 and 358 000, and is nearer to 358 000.

17 (D) Note (10 - 1) x 4 = 9 x 4 = 36; 80 ÷ 2 - 4 = 40 - 4 = 36;
3 x 8 + 2 x 6 = 24 + 12 = 36; 8 + 4 x 3 = 8 + 12 = 20

18 (C) 1m = 100cm, 1m^2 = 100 x 100cm^2 = 10 000cm^2

19 (B) 7km costs 89c in fuel
i.e. 1km costs $\frac{89}{7}$c
Thus 203km costs $\frac{89}{7}$ x 203c ÷ $\frac{91}{7}$ x 200c = 13 x 200c = $26.60

20 (D) Value = 3 + 7 x (7 - 3) = 3 + 7 x 4 = 3 + 28 = 31

21 (C) $^{1620}/_{360} = {}^{162}/_{36} = {}^{81}/_{18} = {}^{9}/_{2} = 4.5$. No. of complete revolutions = 4.

22 (C) **23 (A)** No. cubes = $\dfrac{6 \times 10 \times 8}{2 \times 2 \times 2} = 3 \times 5 \times 4 = 60$

24 (B) Cost of 4½h stay = cost of 5h stay
i.e. cost = $10 + 4 x $5 = $30

25 (A)

26 (C)

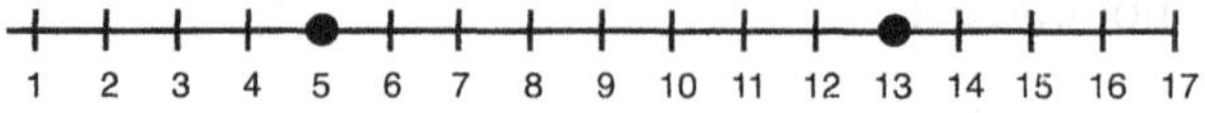

Try each possible answer for number line above.
If number is 1, then distance from 1 to 13 is 12 units and from 1 to 5 is 4 units. This is so since 12 = 3 x 4. If number is 11, then distance to 13 is 2 units and to 5 is 6 units. This is not correct since $2 \neq 3 \times 6$, and so on.

27 (D) Each hour, the minute hand moves through 360° and in 15 minutes through 90°.

28 (D) From 2p.m. to 11a.m. next day is 21h, and clock loses 21 x 3 = 63 minutes.
Clock will read 11a.m. - 63 minutes = 9:57a.m.

29 (B) The consecutive odd numbers starting from 1 are 1, 3, 5, 7, 9, 11, …
The sum of the first 2 numbers is 4; first 3 is 9; first 4 is 16, first 5 is 25; first 6 is 36 and so on.
Noting $\sqrt{4} = 2$, $\sqrt{9} = 3$, $\sqrt{16} = 4$, $\sqrt{25} = 5$, $\sqrt{36} = 6$ then if the sum is 576, the number of odd numbers needed is $\sqrt{576} = 24$, since $24^2 = 24 \times 24 = 576$.

30 (C) From the table, the sums 2, 3, 4, 5, 6, 7, 8, 9, 10, 11, 12 occur respectively 1, 2, 3, 4, 5, 6, 5, 4, 3, 2, 1 times. The most common sum is 7, i.e. the most likely sum is 7.

31 (B) Weight of 5 boys = 5 x 70 = 350kg
Weight of 4 girls = 4 x 61 = 244kg
Total weight of 9 children = (350 + 244) = 594kg
Average weight of 9 persons = 594 ÷ 9 = 66kg

32 (D) A number is divisible by 6 if it is divisible by both 2 and 3, i.e. if it is even and the sum of the digits is divisible by 3. A number is divisible by 9 if the sum of the digits is divisible by 9. We try each pair of possible answers.
Thus, taking 288 and 270 we see they are divisible by 6 and 9, but in 270 the units digit 0 is not greater than the tens digit 7. This pair is not correct. Continue for the other pairs.

33 (C) By trial, we see that the least number of colours is 3.

34 (B) If we cut out the net and fold to form a cube, keeping the faces as given, we see that if R is on the base, then Q is on the face to the left of R and P is at the top of the cube, opposite R. Faces S, T are respectively at the back & front of the cube whilst face U wraps around to complete it, and is opposite face Q.

35 (A) The possible totals using
1 coin is 5c, 10c, 20c, 50c
2 coins is 15c, 25c, 55c, 30c, 60c, 70c
3 coins is 35c, 65c, 75c, 80c
4 coins is 85c
Total number of amounts is (4 + 6 + 4 + 1) = 15

36 (D) Total bill for 18 persons = 18 x \$21 = \$378
Each of 14 persons pays \$378 ÷ 14 = \$27; that is, each pays (\$27 - \$21) = \$6 extra.

37 (C) By counting, number of small squares along perimeter of chessboard is (8 + 7 + 7 + 6) = 28, starting from corner shown.

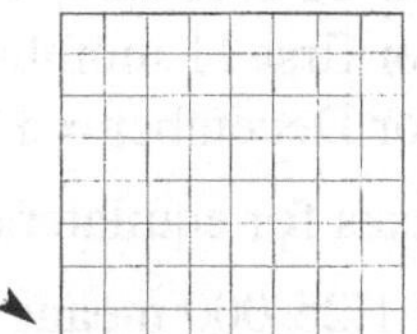

38 (B) The addition is too much by (35 095 - 35.95) i.e. 35 059.05. To correct this, we need to subtract 35 059.05.

39 (D) If P, Q, R, S stand for Peter, Quincey, Raymond and Susie respectively then possible arrangements in a line are listed, so that R, S are together.
PQRS, PQSR, PRSQ, PSRQ, RSPQ, SRPQ
QPRS, QPSR, QRSP, QSRP, RSQP, SRQP

40 (C) Even numbers with digit 2 are
2, 12, 20, 22, 24, 26, 28, 32, 42, 52, 62, 72, 82, 92, 102,112, 120, 122, 124
The digit occurs 21 times.

41 (D) Simplest is roll a coin on a table.

42 (D) The shaded triangle is moved inside the square keeping the right angle at successive corners, in an anti-clockwise direction.

43 (D) Try each possible answer, forming a table
Thus, taking Kate as 25 years, we obtain line 1.
This is incorrect since Angela is 14 not 24.
Continue with other trials.

Angela	Kate	Tony
24	25	19
	3	

44 (C) Try each possible answer.
Thus if there were 10 students, then if Janine is number 6 from left, Janine is number 5 from right. Try again.

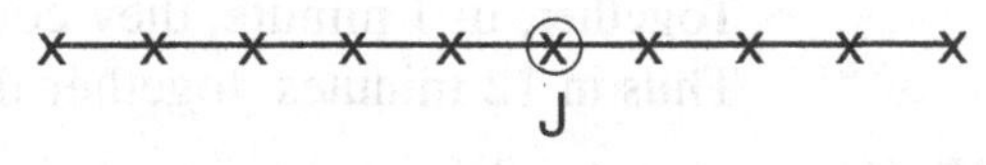

45 (C) Note sequence 3, 7, 15, 31, … is (4 -1), (8 - 1), (16 - 1), (32 - 1), …
The next term is 64 - 1 = 63.

PAPER 9

1 (D) Try each possible answer.

2 (D) Remaining angle = 360° - (50° + 80° + 110°) = 120°
Percentage spent on food = $\frac{120}{360}$ x 100% = $33\frac{1}{3}$%

3 (C) Amount spent on food = $\frac{120}{360}$ x \$360 = \$120

4 (D) Arc ABC = $\frac{6}{30}$ i.e. $\frac{1}{5}$ of circumference
Angle AOC = $\frac{1}{5}$ of complete angle at O = $\frac{1}{5}$ of 360° = 72°

5 (B) Area paddock = 0.24ha = 0.24 x 10 000m^2 = 2 400m^2
Shortest side = A ÷ L = 2 400 ÷ 60 = 40m
Perimeter = 2 (L + B) = 2 (60 + 40) = 200m

6 (C) 1km = 1 000m = 1 000 x 1 000 = 1 000 000mm

$$\text{Percentage} = \frac{247}{1\,000\,000} \times 100\% = \frac{247}{10\,000}\% = 0.0247\%$$

7 (C) Value = 0.04 + 0.3 = 0.34

8 (C) Salary for 12 months = 12 x $1 010 = $12 120
Salary for first 11 months = 11 x $1 000 = $11 000
Salary for December = $12 120 - $11 000 = $1 120

9 (B) No. of axes for equilateral triangle is 3 and for rectangle is 2. Difference = 3 -2 = 1

10 (D) Scale of 1:25 000 means that 1cm on map represents a distance of 25 000cm = 250m
i.e. 250m distance is represented by 1cm on map
Thus 1m distance is represented by $\frac{1}{250}$cm on map
i.e. 2 000m distance is represented by $\frac{1}{250}$ x 2 000 = 8cm on map.

11 (D) Increase in cost = $\frac{1}{3}$ x $72.60 = $24.20
New price = $72.60 + $24.20 = $96.80

12 (D) Try possible answers, omitting $36 as it is obviously too big, since selling price is $30.
Thus if cost price is $24, increase = $\frac{1}{5}$ x $24 = $4.80 and selling price ≠ $30 and so on.

13 (D) 0.75 of 1km = $\frac{3}{4}$ x 1 000m = 750m

14 (C) Rates = 65 000 x 2.3c = $ $\frac{65\,000 \times 2.3}{100}$ = $65 x 23 = $1 495

15 (A) Shaded triangle and rectangle have same base AB and same height h.
Area Δ ABE = ½b x h
Area rectangle ABCD = b x h
Thus shaded triangle is ½ rectangle in area.

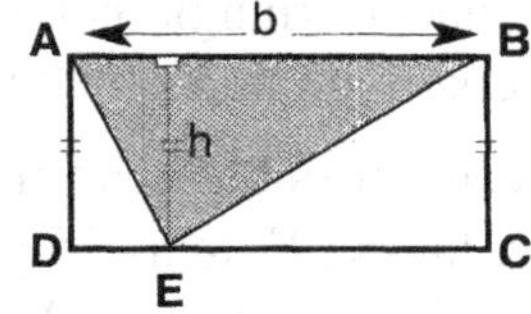

16 (C) In 1 minute, first tap can fill $\frac{1}{20}$ of bucket and second tap can fill $\frac{1}{30}$ of bucket.
Together, in 1 minute, they could fill ($\frac{1}{20}$ + $\frac{1}{30}$) = $\frac{3+2}{60}$ = $\frac{5}{60}$ = $\frac{1}{12}$ of bucket.
Thus in 12 minutes, together they would fill the bucket.

17 (C) 15 ÷ 4 = 3¾; need to buy 4 tins

18 (B) **19 (A)** **20 (A)**

21 (C) Since 6 is divisible by 2, we need the smallest number divisible by 5, 6 i.e. 30

22 (A) If breadth is n cm, then length is 5 x n cm.
Area = 5 x n x n cm^2, but area = 80cm^2
Thus 5 x n x n = 80, i.e. n x n = 80 ÷ 5 = 16
Then n = 4 and perimeter is 2(20 + 4) = 48cm.

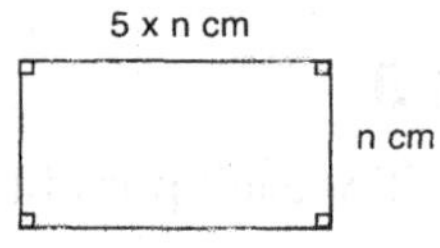

23 (B) If square has perimeter 48cm, its side is 48 ÷ 4 = 12cm and area is 12^2 = 144cm^2.
Difference in areas = (144 - 80) = 64cm^2.

24 (B) $\frac{0.1 \times 0.4}{5} = \frac{0.04}{5}$ = 0.008   5) 0.040 = 0.008

25 (C) Nets which will not form an open box are as follows:
(ii) Two squares on same side of 3 squares will cover the same face of cube, leaving 2 open faces.
(iii) First and last squares will overlap leaving 2 open faces.
(vi) There are 4, not 5 squares.

26 (C) Try each possible answer.

27 (A) Minute hand rotates through a complete revolution once every hour, i.e. 24 times per day. As June has 30 days, no. of rotations = 30 x 24 = 720.

28 (A) \$10 = 1 000c; no. of stamps = 1 000 ÷ 45 = 22 rem 10
Thus only 22 stamps can be purchased for \$10.

29 (A) Distance PQ = speed x time = 80 x ½ = 40km.
Distance Julie returns = 60 x ⅓ = 20km (to R).
She is then (40 - 20) = 20km north of P.

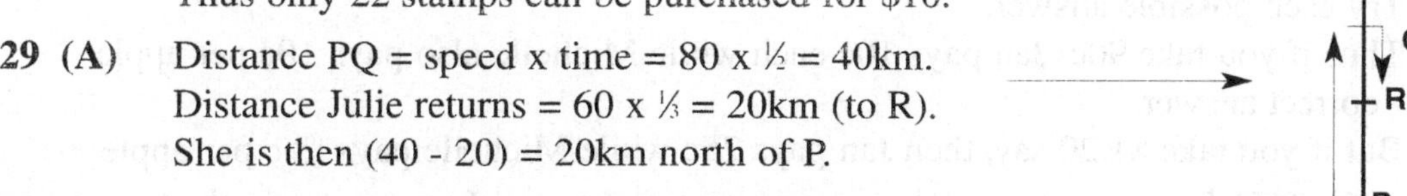

30 (A) Vertical cut through centre O shown

O

31 (D) Try each possible answer.

32 (C) In the sequence 2, 3, 7, 16, 32, … the difference between the terms is 1, 4, 9, 16, … i.e. 1^2, 2^2, 3^2, 4^2, …. The next term in the sequence is 32 + 5^2 = 32 + 25 = 57.

33 (D) Find a pattern by forming a table thus:
Note the perimeter in cm is 2 more than the number of triangles.
Thus perimeter of 20 such triangles is (20 + 2) = 22cm.

No. of Δs	Perimeter
1	3cm
2	4cm
3	5cm
4	6cm
5	7cm

34 (B) The cube has 6 faces; and the smallest even number is 22.
To obtain the largest consecutive even number we take 22, 24, 26, 28, 30, 32 for the faces.

35 (A) If the side of the square sheet is a cm, then the perimeter of a half sheet is (a + ½a + a + ½a) = 3a cm. Since this perimeter is 18cm, then 3a = 18, i.e. a = 6, and thus the original square has perimeter 4 x 6 = 24cm.

a
½a
½a
a

36 (B) We try to form a pattern linking the position in the row of the shape and the number of blocks in the shape, and form a table thus:
Note the number of blocks increases by 3, i.e. the number of blocks in the 4th shape is 10. We want to relate the position S of the shape and the number N of blocks.
After some thought, we should see that N = 3 x S - 2;
check 1 = 3 x 1 - 2; 4 = 3 x 2 - 2; 7 = 3 x 3 - 2; 10 = 3 x 4 - 2 and so on.
The number of blocks for the 10th shape is thus 3 x 10 - 2 i.e. 28.

Shape S	No. of Blocks N
1	1
2	4
3	7
4	10

37 (D) Sketches will help here.

A triangular cut for the rectangular prism is shown.

A
B

rectangular prism | triangular prism | square pyramid | triangular pyramid

38 (B) Remember the first cut makes 2 pieces, i.e. Ben needs 2 cuts to make 3 pieces. Time for this is 3 minutes, i.e. each cut takes 3 ÷2 = 1½ minutes.
To obtain 9 pieces, he needs 8 cuts which will take 8 x 1½ = 12 minutes.

39 (C) Easily seen if take whole numbers such as 5, 2 say; here 5 + 2, 5 x 2 are whole numbers always but 5 ÷ 2 is not a whole number. (Although if we take say 8, 2 then 8 ÷ 2 is a whole number).

40 (D) 100cm = 1m, i.e. 150cm = $1\frac{1}{2}$ = 1.5m, 50cm = 0.5m
Area = 1.5 x 0.5 = 0.75m^2

41 (B)

42 (A) Try each possible answer.
Thus if you take 90c, Jan pays 10c each while Michelle also pays 10c per apple - correct answer.
But if you take $1.20 say, then Jan pays 25c while Michelle pays $\frac{100}{7}$c per apple - not correct.

43 (A) Find the length systematically to avoid errors.
Sum of horizontal lengths = 5 x 10m = 50m
Sum of vertical lengths = 5 x 10m = 50m
Total length = (50 + 50) = 100m

44 (B) Volume of water displaced is equal to volume of block dropped into tank.
This volume = 6 x 3 x 2 = 36cm^3
Amount of water displaced = 36mL, since 1cm^3 = 1 mL

45 (A) Each face of each cube is of area 1cm^2. By counting
Solid P has surface area (2 x 8 + 2 x 4 + 2 x 2) = 28cm^2
Solid Q has surface area 6 x 4 = 24cm^2
Solid R has surface area (10 x 2 + 2 x 4) = 28cm^2
Solid S has surface area (2 x 8 + 2 x 3 + 2 x 4 + 2 x 2) = 34cm^2
Difference required = (34 - 24) = 10cm^2

PAPER 10

1 (C) Note 2^4= 2 x 2 x 2 x 2 = 16 ; 4^2 = 4 x 4 = 16. Value = 16 + 16 = 32

2 (B) If 3/5 of a number is 42
then 1/5 of the number is 42 ÷ 3 = 14
and 5/5 i.e. the number is 14 x 5 = 70

3 (D) Write all as fractions with denominator 20 i.e. $\frac{3}{5} = \frac{12}{20}$, $\frac{1}{2} = \frac{10}{20}$, $\frac{9}{20}$, $\frac{3}{4} = \frac{15}{20}$
or as decimals $\frac{3}{5} = \frac{6}{10}$ = 0.6, $\frac{1}{2}$ = 0.5, $\frac{9}{20} = \frac{45}{100}$ = 0.45, $\frac{3}{4}$ = 0.75

4 (D) **5 (D)** **6 (A)** **7 (C)** Try each possible answer.

8 (D) Try each possible answer.
Thus, if □ = 4, L.H.S. = 4 x 8 = 32; R.H.S. = 2 x (4 + 2) + 14 = 26

9 (A) 1 620c buys 12L oil
i.e. 1c buys $\frac{12}{1620}$L oil
and 405c buys $\frac{12}{1620}$ x 405 = $\frac{12}{4}$ = 3L oil

10 (D) In 4, 7, 12, 19, 28, □ the differences between the terms are 3, 5, 7, 9 respectively.
The value of □ is 28 + 11 = 39

11 (B) We work out the costs for the same quantity of washing powder.
Noting 2 x 375g = 750g and 4 x 375g = 1 500g = 1.5kg and 6 x 250g = 1 500g = 1.5kg we work the cost of 1.5kg of each powder.
At 250g for $6.25, 1.5kg costs 6 x $6.25 = $37.50
At 375g for $9.00, 1.5kg costs 4 x $9.00 = $36.00
At 1kg for $25.50, 1.5kg costs $1\frac{1}{2}$ x $25.50 = $38.25
Last item, 1.5kg costs $37.65
Best buy is now easily seen.

12 (B) Original number 312.4, new number 412.3. Difference = 412.3 - 312.4 = 99.9

13 (A) Cost books \$4; left \$16; cost bag \$12, left \$4

14 (D) $(\frac{3}{4} + \frac{1}{2}) \times 4 = (1\frac{1}{4}) \times 4 = 5$

15 (A) $\frac{1}{2} \times (3\frac{1}{2} \times 4) = \frac{1}{2} \times (14) = 7$

16 (C) $\frac{7 \times 2}{5 \times 7} = \frac{2}{5}$; $\frac{(2 + 4)}{(7 + 8)} = \frac{6}{15} = \frac{2}{5}$; $\frac{(2 + 7)}{(5 + 7)} = \frac{9}{12} = \frac{3}{4}$;

$(6 + 2) \div (11 + 3 \times 3) = 8 \div (11 + 9) = 8 \div 20 = \frac{2}{5}$

17 (B) $4\frac{1}{2} - 3\frac{1}{4} = (4-3) + (\frac{1}{2} - \frac{1}{4}) = 1\frac{1}{4}$

18 (A) Note $1\text{m}^3 = 1\ 000\text{L}$, i.e. $\frac{1}{2}\text{m}^3 = 500\text{L}$
OR As 1m = 100cm, then $1\text{m}^3 = 100 \times 100 \times 100 = 1\ 000\ 000\text{cm}^3$
i.e. $1\text{m}^3 = 1\ 000\ 000\text{mL} = \frac{1\ 000\ 000}{1\ 000}\ \text{L} = 1\ 000\text{L}$. Thus $\frac{1}{2}\ \text{m}^3 = 500\text{L}$.

19 (B) Note theoretically, on every bounce the ball goes up and down in the same vertical line. The sketch is drawn to help understand the solution.

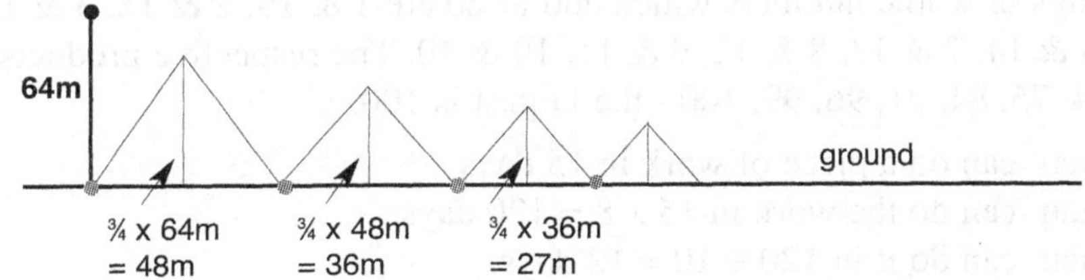

Distance travelled = (64 + 2 x 48 + 2 x 36 + 2 x 27) = 286m

20 (D) The numerators of the fractions decrease respectively by 12, 10, 8, 6 and the denominators decrease by 1.
The next term will be $\frac{7 - 4}{3 - 1} = \frac{3}{2}$

21 (A) **22 (D)** **23 (C)** He receives 125 x 7 c = 875c = \$8.75

24 (D) Cost = 200 x \$2.93 = 2 x \$293 = \$586

25 (B) Note a + c = 12 and b + d = 7
Perimeter= 12 + 7 + a + b + c + d
= 19 + (a + c) + (b + d) = 19 + 12 + 7 = 38cm

26 (D) Try the possible answers
Thus if X = 3, Y = 7 number is 2 372, which leaves remainder 4 when divided by 8, and so on.
8) 2372
296 rem 4

27 (C) All radii of the circle are 8cm. Length = 4 x 8 = 32cm

28 (A) Break into 2 rectangles as shown.
Area of figure = 11 x 9 - 4 x 2 = 91 sq. units

29 (C) 147 681 lies between 147 000 and 148 000, and is nearer to 148 000

30 (A) 2 500 ÷ 47 = 53 rem 9
÷ 53

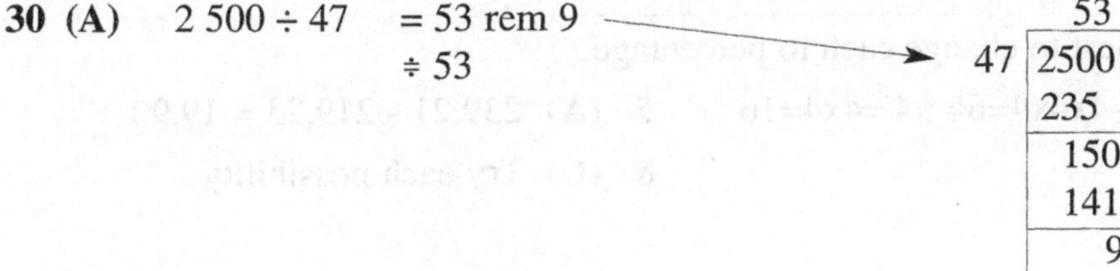

31 (B) No. of tiles $= 9\frac{1}{3} \div 1\frac{1}{3} = \frac{28}{3} \div \frac{4}{3} = 28 \div 4 = 7$

32 (C) Amounts shown on graph: Rent \$120, Food \$60, Clothes \$40, Savings \$30; i.e. her total salary is \$250.
Fraction of salary spent on rent $= \frac{120}{250} = \frac{12}{25}$

33 (C) Percentage saved $= \frac{30}{250} \times 100\% = 12\ \%$

34 (A) Fraction of salary spent on food $= \frac{60}{250} = \frac{6}{25}$
Angle at centre of circle for food section of a pie chart
$= \frac{60}{250}$ of $360° = \frac{432°}{5} = 86\frac{2}{5}° = 86°$, to nearest°

35 (D) Now 648 x 33 = (2 x 324) x (3 x 11) = 2 x 3 x (324 x 11)
= 6 x 3564, we are given 324 x 11 = 3564
Thus, we need to multiply 3564 by 6.

36 (A) Double the top number and subtract 1 to get the corresponding bottom number; i.e. 2 x 2 - 1 = 3, 2 x 9 -1 =17, 2 x 4 - 1 = 7, 2 x 12 - 1 = 23.
Likely number is 2 x 7 - 1 = 13.

37 (A) Total = \$118.48; average = \$118.48 ÷ 4 = \$29.62

38 (A) The pairings of whole numbers which add to 20 are 1 & 19, 2 & 18, 3 & 17, 4 & 16, 5 & 15, 6 & 14, 7 & 13, 8 & 12, 9 & 11, 10 & 10. The respective products are 19, 36, 51, 64, 75, 84, 91, 96, 99, 100 - the largest is 100.

39 (A) If 8 men can do a piece of work in 15 days
then 1 man can do the work in 15 x 8 = 120 days
and 10 men can do it in 120 ÷ 10 = 12 days.

40 (C) 4m/s is equivalent to 4 x 60m/minute
This is equivalent to 4 x 60 x 60 m/hour
That is, to $\frac{4 \times 60 \times 60}{1\,000} = \frac{144}{10} = 14.4$km/h

41 (D) If 2□7 x 39 = 11 193 then 2□7 = 11 193 ÷ 39
We divide 11 193 by 39 and see that →
11 193 ÷ 93 = 287, i.e. □ = 8

```
      287
39 | 11 193
     78
     339
     312
      273
      273
```

42 (D) The numbers in opposite sectors are such that the larger is the square of the smaller, e.g. $64 = 8^2$, $25 = 5^2$, $9 = 3^2$
The missing number must be $13^2 = 169$

43 (B) Note diagrams **(A)**, **(C)**, **(D)**, **(E)** can be obtained from one another by rotation in the plane of the paper, but diagram **(B)** cannot be obtained thus, {The other diagrams must be flipped over to obtain diagram **(B)**}.

44 (C) One axis of symmetry only →

45 (D) Try each possible answer.
Thus if the number is 18, then its square $18^2 = 324$ added to twice 18 gives 324 + 36 = 360. This is not the required answer.

PAPER 11

1 (A) **2 (D)** Best to change each to percentage

3 (B) **4 (C)** 4^3-4x4x4=64 ; 4^2=4x4=16 **5 (A)** 239.21 - 219.23 = 19.98

6 (A) **7 (A)** **8 (C)** Try each possibility

9 (B)

Day	1	2	3	4	5
Me	\$81	\$162	\$324	\$648	\$1 296
Fred	\$16	\$48	\$144	\$432	\$1 296

10 (B) 9kg cheese cost \$64.17
i.e. 1kg cheese cost \$64.17 ÷ 9 = \$7.13
and 11kg cheese cost \$7.13 x 11 = \$78.43

11 (A) Largest 9741, smallest 1479; difference 8262

12 (A) In sequence 1, 6, 13, 22, 33, □ the difference in the terms is 5, 7, 9, 11.
Next term is 33 + 13 = 46

13 (D) Find cost per litre for each, noting 4 x 250mL = 1L, 5 x 600mL = 3L, 2 x 2.5L = 5L.
Cost 250mL for 34c, then cost 1L = 4 x 34c = 136c
Cost 600mL for 84c, then cost 3L = 5 x 84c = 420c and
thus cost 1L = 420c ÷ 3 = 140c
Cost 2.5L for 335c, then cost 5L = 2 x 335c = 670c and
thus cost 1L = 670c ÷ 5 = 134c
It is easy to decide correct answer now.

14 (D) $\frac{1}{2}(\frac{1}{2} + \frac{3}{4}) = \frac{1}{2}(\frac{5}{4}) = \frac{5}{8}$

15 (D) **16 (D)** Values of parts in order are 10, 10, 10, 5 **17 (A)**

18 (B) **19 (C)** **20 (A)** **21 (A)** **22 (C)**

23 (A) Sequence 1, $\frac{7}{8}$, $\frac{3}{4}$, $\frac{5}{8}$, $\frac{1}{2}$, □ can be written as $\frac{8}{8}$, $\frac{7}{8}$, $\frac{6}{8}$, $\frac{5}{8}$, $\frac{4}{8}$, □. Thus □ = $\frac{3}{8}$.

24 (A) Try each possible answer.

25 (D) In one year's time, i.e. next year, the sum of my age and my brother's age is 35 years.
Thus 7 + 1 = 8 years earlier, the sum of these ages must have been (35 - 2 x 8) = 19 years

26 (B) Draw in broken lines and insert letters as shown.
Now a + 4 + b = 17 and c = 5 + 7 = 12
Perimeter of figure = 17 + c + a + 5 + 4 + 2 + b + 9
= 17 + c + (a + 4 + b) + 5 + 2 + 9
= 17 + 12 + 17 + 16 = 62cm

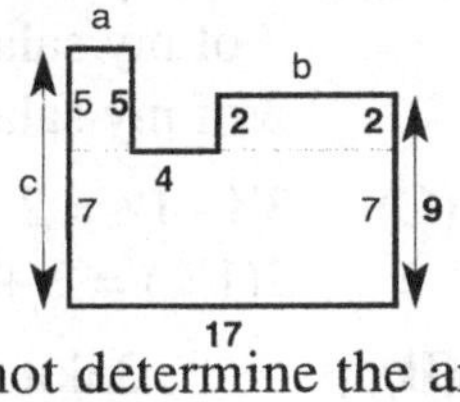

27 (D) Area = 17 x 7 + 5a + 2b , and as we do not know a, b, we cannot determine the area.

28 (D) **29 (C)** Total no. cars 30; percentage = $\frac{6}{30}$ x 100 = 20%

30 (A) Fraction = $\frac{16}{30} = \frac{8}{15}$ **31 (D)** Angle = $\frac{8}{30}$ of 360° = 96°

32 (A) Length = $\frac{4}{30}$ of 15cm = 2cm **33 (C)** Try each possibility

34 (B) Distance = 80 x 9 = 720km; time = 720 ÷ 90 = 8h

35 (C) $\frac{636}{4} = \frac{636}{4} \times \frac{3}{3} = \frac{636}{12} \times 3 = 53 \times 3$ as $\frac{636}{12} = 53$
Thus we need to multiply 53 x 3

36 (B) If 5% of a number is 94
then 100% of number is 20 x 94 = 1880

37 (D) Try each possibility

38 (D) The bottom numbers in the table are 3 times the corresponding top numbers plus 1;
e.g. 10 = 3 x 3 +1, 25 = 3 x 8 + 1, 16 = 3 x 5 + 1 etc. Missing number = 3 x 9 + 1 = 28

39 (A) **40 (C)** 46 351 is between 46 300 and 46 400

41 (A) If the 2 digit number formed by the tens and units digit of the larger number is greater than the corresponding 2 digit number of the smaller number, then the digit under the hundreds digit is 0, and if greater is replaced by 'smaller' in the above, then the digit under the hundreds digit is 9.

e.g. 9482 note 82>67 OR 9467 note 67 <82

9482	note 82>67	9467	note 67 <82
2467 -		2482 -	
7015		6985	

42 (A) Sum of numbers in left to right diagonal is 84; then missing number in first column is 12 and then x = 34

43 (B) 240c buys 3kg flour
40c buys 3kg ÷6 = $\frac{1}{2}$ kg flour

44 (C) Width = $\frac{1}{2}$ (perimeter - 2 x length) = $\frac{1}{2}$ (180 - 2 x 70) = 20cm

45 (B) Remove 2 adjoining matches at middle of square.

PAPER 12

1 (B) 0.05 + 0.5 = 0.55 **2 (B)**

3 (D) No. is $\frac{1}{2}$(74.81 + 84.37) = $\frac{1}{2}$(159.18) = 79.59

4 (C) 5^3=5x5x5=125; 5^2=5x5=25 **5 (A)** Change each to a percentage.

6 (B) Using letters A, B, C, D to represent Ally, Babs, Candy, Don respectively the different arrangements with A, B, C together are:
ABCD, ACBD, BACD, BCAD, CABD, CBAD, DABC, DACB, DBAC, DBCA, DCAB, DCBA.

7 (B) **8 (C)** **9 (A)**

10 (B) $\frac{2}{5}$ of my salary is $140 (Note 1 - $\frac{3}{5} = \frac{2}{5}$)
$\frac{1}{5}$ of my salary is $140 ÷ 2 = $70
$\frac{3}{5}$ of my salary is $70 x 3 = $210; this is rent

11 (C) $3\frac{1}{4} - 1\frac{1}{2} = 2 + \frac{1}{4} - \frac{1}{2} = 2 + \frac{1-2}{4} = 1 + \frac{5-2}{4} = 1\frac{3}{4}$
$2(1\frac{3}{4}) = 2 + 1\frac{1}{2} = 3\frac{1}{2}$

12 (B) 35km uses 5 L of fuel
i.e. 1km uses $\frac{5}{35} = \frac{1}{7}$ L fuel
Thus 42km uses $\frac{1}{7}$ x 42 = 6 L fuel

13 (C) Calculate cost per kg, noting 4 x 250g = 1kg, 5 x 600g = 3kg, 5 x 1.2kg = 6kg
2 x 2.5kg = 5kg

250g for $1.28 means	1kg for 4 x $1.28 = $5.12
600g for $3.12 means	3kg for 5 x $3.12 = $15.60, and
thus	1kg for $15.60 ÷ 3 = $5.20
1.2kg for $6 means	6kg for 5 x $6 = $30, and
thus	1kg for $30 ÷ 6 = $5.00
2.5kg for $13.25 means	5kg for 2 x $13.25 = $26.50
and thus	1kg for $26.50 ÷ 5 = $5.30

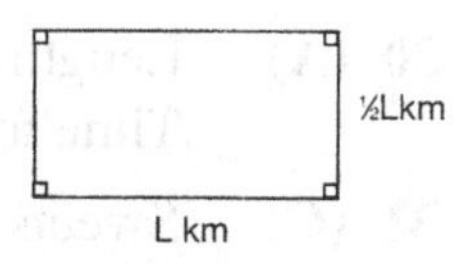

14 (D) If length is L km, then width is $\frac{1}{2}$L km.
Perimeter is (L + $\frac{1}{2}$L + L + $\frac{1}{2}$L) = 3L km.
But by data, perimeter is 3km; thus L = 1km, and breadth is $\frac{1}{2}$km.
Area = L x B = 1 x $\frac{1}{2}$ = $\frac{1}{2}$km^2

15 (B) 7) 29 = 4 rem1 7) 135 = 19 rem2 First multiple of 7 between 29 and 135 is 7 x 5 = 35 and last multiple is 7 x 19 = 133.
There are 19 - 4 = 15 multiples of 7

16 (B) $4\frac{1}{2} \div 3 = 1\frac{1}{2}$; $1\frac{1}{2} \times 2 = 3$

17 (C) 24 x 24 = 24^2, a perfect square; 24 x 12 = (2 x 12) x 12 = 2 x 12^2, not a perfect square; 24 x 6 = (4 x 6) x 6 = 2^2 x 6^2, a perfect square; 24 x 18 = (4 x 6) x (3 x 6) = 2^2 x 6^2 x 3, not a perfect square.

18 (A) **19 (C)**

20 (A) I save or spend on stamps ($\frac{1}{2}$ + $\frac{1}{5}$) = $\frac{7}{10}$ of my pocket money; this leaves $\frac{3}{10}$.
Thus $\frac{3}{10}$ of my pocket money = $3.21
i.e. $\frac{1}{10}$ of my pocket money = $3.21 ÷ 3 = $1.07
and $\frac{2}{10}$ = $\frac{1}{5}$ of my pocket money = $1.07 x 2 = $2.14
This is the amount I spent on stamps.

21 (D) 100% - 30% = 70%
Amount left after tax = 70% of $40 000 = $\frac{70}{100}$ x $40 000 = $28 000

22 (A)

23 (D) The sequence is $\frac{1}{1}$, $\frac{3}{2}$, $\frac{9}{4}$, $\frac{27}{8}$, □. Each numerator is obtained by multiplying the previous one by 3, and each denominator is obtained by multiplying the previous one by 2.
The next number is $\frac{27 \times 3}{8 \times 2} = \frac{81}{16}$

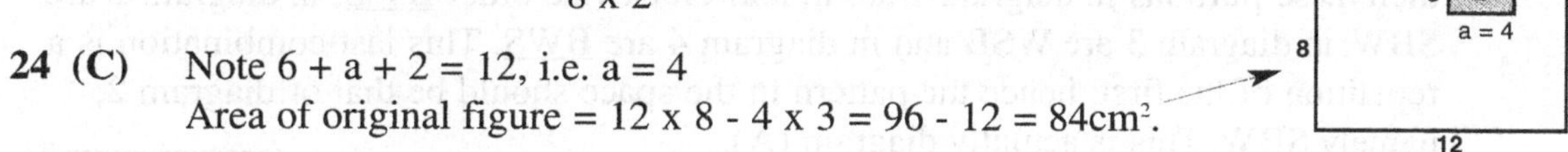

24 (C) Note 6 + a + 2 = 12, i.e. a = 4
Area of original figure = 12 x 8 - 4 x 3 = 96 - 12 = 84cm^2.

25 (B) Note a + b + c + d = 10 and p + q + r + s = 14
Perimeter = 10 + 14 + (a + b + c + d) + (p + q + r + s)
= 10 + 14 + 10 +14 = 48cm

26 (A) Angle at centre of circle is 360°
Fraction owning IBM = $\frac{135}{360} = \frac{27}{72} = \frac{3}{8}$

27 (B) Angle at centre for Mac = 360° - (135° + 90°) = 135°
Angle at centre for IBM and Mac = = 135° + 135° = 270°
Percentage required = $\frac{270}{360}$ x 100% = 75%

28 (B) Angle at centre for Dells = 90° - 45° = 45°
That is, angle of 45° represents 4 persons
i.e. angle of 360° = 8 x 45° represent 4 x 8 = 32 persons
No. of people in class is 32.

29 (C) Try the possibilities for □.
Thus if □ = 0, is 1071 a multiple of 7? Yes 7) 1071 = 153
if □ = 9, is 1971 a multiple of 7? No 7) 1971 = 281 rem 4
Then try possibility (B) and so on.

30 (A) Length of journey = 48 x 3 = 144km
Time to do journey at 36km/h = 144 ÷ 36 = 4h

31 (C) Sweets cost ¼ x \$10 = \$2.50; goldfish cost ½ x \$10 = \$5
Amount left = \$10 - (\$2.50 + \$5) = \$2.50

32 (B) Square of top number + 1 gives corresponding bottom number;
e.g. $2^2 + 1 = 5$, $7^2 + 1 = 50$, $4^2 + 1 = 17$. Answer $9^2 + 1 = 82$

33 (D) Average = ¼(17.80) = 4.45

34 (C) By multiplication 33 x 33 = 1089; this lies in hundreds between ten hundred (i.e. 1 000) and eleven hundred (i.e. 1 100), and is nearer to 1 100.

35 (D) Note all prime numbers, except 2, are odd. Quadrilaterals can have 3 obtuse angles, e.g. these could be say 95°, 105°, 110° and then remaining angle is 50° since the angle sum of a quadrilateral is 360°. This 'average' idea is easy to illustrate, for example for set 2, 3, 6, 9 the average must be larger than 2 and smaller than 9. In the triangle any two sides together are greater, not less, than the remaining side.

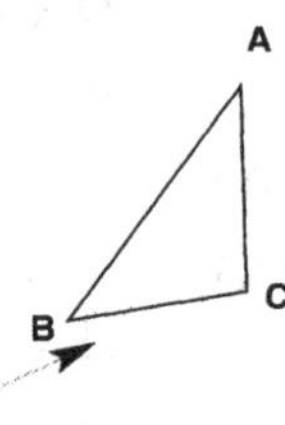

36 (B) Consider semicircles, each consisting of 4 sectors, beginning with 6, 8 , 10, 12 and proceeding 1 sector at a time in a clockwise direction, i.e. the next semicircle is 8,10, 12, 16; the next is 10, 12, 16, blank; the next 12, 16, blank, 24; the last 16, blank, 24, 32.
Note in each, the last number is twice the first e.g. 12 = 2 x 6, 16 = 2 x 8, 24 = 2 x 12, 32 = 2 x 16. Thus blank = 2 x 10 = 20.

37 (A) Each of the 4 given diagrams consists of 3 distinct portions; namely a long L, a straight part and a short L. If we use B, W, S for black, white, spotted respectively then these portions in diagram 1 are in anti-clockwise order <u>BWS</u>, in diagram 2 are SBW, in diagram 3 are WSB and in diagram 4 are <u>BWS</u>. This last combination is a repetition of the first; hence the pattern in the space should be that of diagram 2, namely SBW. This is actually diagram (A).

38 (B) Each of the 8 triangles is the same size and hence the large square is twice the area of the small square. →

39 (C) Let S be sum of counting numbers 1, 2, 3, …, 50

Thus	S = 1 + 2 + 3 + …… + 49 + 50
Reversing,	S = 50 + 49 + 48 + …… + 2 + 1
Adding in pairs	2S = (1 + 50) + (2 + 49) + (3 + 48) + …+ (49 + 2) + (50 + 1)
ie	2S = 50 pairs each totalling 51, i.e. 50 x 51
Thus	$S = \frac{50 \times 51}{2}$ = 25 x 51 = 1275, multiplying

That is, sum is nearest to 1 200 of given possibilities.
Note: This method of summing is often useful.

40 (B) The factors of 144 are 1 x 144, 2 x 72, 3 x 48, 4 x 36, 6 x 24, 8 x 18, 9 x 16, 12 x 12. The required pair is 8, 18, since their sum is 26. The larger exceeds the smaller by 10.

41 (A) A diagram is essential here.
From Home (H), the girl proceeds respectively to A, B, C, D, E, F. (Note the calculations on the journey).
Thus she is at F, which is (7 + 2 + 3) = 12n miles due south of H. To reach H, she must sail 12n miles due north.

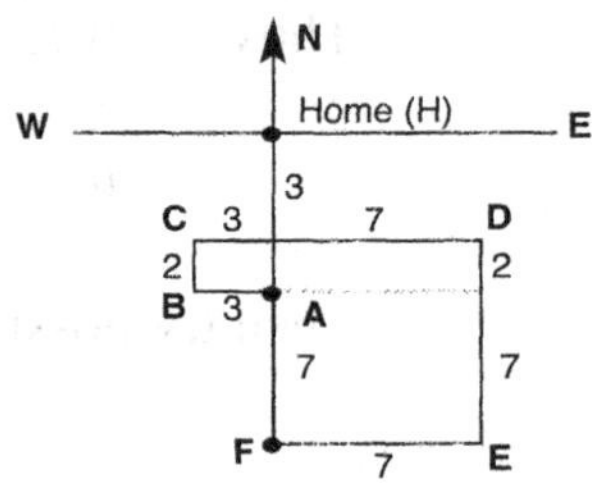

42 (A) If it is 11:54p.m. Monday in Perth, then it is (11:54p.m. + 2h) = 1:54a.m. Tuesday in Sydney.

43 (B) For 10cm strip on road, cost is $70
For 1cm strip on road, cost is $70 ÷ 10 = $7
For 7cm strip on road, cost is $7 x 7 = $49
Saving is ($70 - $49) = $21.

44 (A) Sides of outside square are 24m
Distance around outside path = 4 x 24 = 96m

24m
20m
20m
24m
2m

45 (A) Pattern of colours of beads is
red, black, white, blue, white, black, | red, black, white, blue, ...|
It repeats after 6 colours. Note 55 = 9 x 6 + 1, then this pattern of 6 beads occurs 9 times and the next bead, i.e. the 55th bead is red.

PAPER 13

1 (C) Numbers divisible by both 2, 5 are divisible by 2 x 5 = 10.
Numbers less than 100 leaving remainder of 1 when divided by 2, 5 are 10 + 1, 20 + 1, 30 + 1, 40 + 1, 50 + 1, 60 + 1, 70 + 1, 80 + 1, 90 + 1,
i.e. 11, 21, 31, 41, 51, 61, 71, 81, 91. Those which are divisible by 3 are 21, 51, 81.

2 (D) Try each possible answer.
Thus if Uncle Bill is 62 years in 6 years time, he is now 56 and thus John is $\frac{1}{6}$ of 56 = $9\frac{1}{3}$.
In 4 years time, Uncle Bill will be 60 and John will be $13\frac{1}{3}$. Note $60 \neq 4 \times 13\frac{1}{3}$.

3 (C) Each student gives (7 - 1) = 6 presents; altogether there will be 7 x 6 = 42 presents.

4 (C) Try each possible answer.
Thus if she gets 20 right, and thus 5 wrong she receives 20 x 4 = 80 marks and loses 5 x 1 = 5 marks, a net total of 80 - 5 = 75 marks.

5 (D) Try each answer.

6 (C) No. of different ways with X as the third letter:
WYXZ, YWXZ, ZYXW, YZXW, WZXY, ZWXY.
There are also 6 ways with X as the fourth letter.

7 (D) No. of mistakes found only by first reader = 70 - 39 = 31
No. of mistakes found only by second reader = 94 - 39 =55
Total no of mistakes = 31 + 55 + 39 = 125

8 (A)

km
0 1 2 3 4 5 6 7 8 9 10

Ben takes 10 ÷ 2 = 5h to swim 10km at 2km/h.
George takes 1h10min to swim from 0 to 2 and float back to 1.
He then takes 1h10min to swim from 1 to 3 and float back to 2.
i.e. at the end of 2h20min he is at the 2km mark.
Similarly, at the end of 3 x 1h10min = 3h30min he is at the 3km mark.
Thus at the end of 8 x 1h10min = 8h80min = 9h20min, he is at the 8km mark.
In the next hour, he swims 2km and has finished his swim (he does not of course float back then).
Thus, George takes 10h20min to reach the 10km mark, a time difference of (10h20min - 5h) = 5h20min longer than Ben.

9 (D) By counting, we see there are (14 + 12 - 1) = 25 rows with (7 + 9 + 1) = 17 seats in each row, a total of 25 x 17 = 425 seats.

10 (D) □ □□ □□□ □□□□ □□□□□

We try to build up a pattern.
1 table can seat 4 x 4 = 16 students
2 tables can seat 6 x 4 = 24 = 16 + 8 students
3 tables can seat 8 x 4 = 32 = 16 + 2 x 8 students
4 tables can seat 10 x 4 = 40 = 16 + 3 x 8 students
5 tables can seat 12 x 4 = 48 = 16 + 4 x 8 students
Thus, 30 tables can seat 16 + 29 x 8 = 248 students

11 (A) If area of smaller floor is L x B, then area larger floor = 2L x 3B = 6 x L x B, i.e. it is 6 times area smaller floor.
Cost for larger floor = 6 x \$120 = \$720

12 (B) Time for first 20km = $\frac{20}{40}$ = $\frac{1}{2}$h, and for second 20km is $\frac{20}{60}$ = $\frac{1}{3}$h
Thus bus takes ($\frac{1}{2}$ + $\frac{1}{3}$) = $\frac{5}{6}$h to go 40km
i.e. it takes $\frac{1}{6}$h to go 40 ÷ 5 = 8km
and hence it takes 1h to go 8 x 6 = 48km, i.e. average speed is 48kmh.

13 (C) Odd numbers possible must be 1,3,5,7,9,11. The combinations which add to 13 are 11,1,1 or 9,3,1 or 7,3,3 or 7,5,1 or 5,5,3. The largest product is 5 x 5 x 3 = 75.

14 (D) If a fraction lies between 0 and 1, the numerator must be less than the denominator. Try some.
If take say $\frac{4}{5}$, then $\frac{4+5}{5+5}$ = $\frac{9}{10}$ Note $\frac{4}{5} < \frac{9}{10}$
If take say $\frac{2}{3}$, then $\frac{2+5}{3+5}$ = $\frac{7}{8}$ Note $\frac{2}{3} < \frac{7}{8}$
If take say $\frac{3}{4}$, then $\frac{3+5}{4+5}$ = $\frac{8}{9}$ Note $\frac{3}{4} < \frac{8}{9}$
We can easily see that options (A), (B), (C) are incorrect and that option (D) is correct.

15 (A) There are (24 - 4) = 20 numbers from 5 to 24, of which 4 are multiples of 5.

16 (D) The faster runner takes 300s to complete 1 lap and thus is at the starting point after 300s, 600s, 900s, 1200s, 1500s, 1800s, 2100s, …
The slower runner takes 1050s to complete 1 lap and is at the starting point after 1050s, 2100s, 3150s, …
From the above, we see that after 2100s they are at the starting point together; the faster runner has completed 7 laps while the slower runner has done only 2 laps.

17 (A) Each of the 9 persons shakes hands with 8 persons, a total of 9 x 8 = 72 handshakes. However, each of these has been counted twice since A shaking B's hand is the same as B shaking A's hand. Thus number of handshakes = 72 ÷ 2 = 36.

18 (C) 30km/h = 30 000m/h = (30 000 ÷ 60)m/min.
= 500m/min.
= 50 000cm/min.
If the car is travelling at 30km/h, then in 1 minute, it moves forward a distance of 50 000cm. In 1 revolution, the wheels travel 100cm. Hence the distance 50 000cm is equivalent to 50 000 ÷ 100 = 500 revolutions of the wheels.
Thus in 1 minute, the wheels are revolving (or spinning) 500 times. That is the wheels are then spinning at a speed of 500 revolutions per minute.

19 (D) Area = 8 x 9 - 7 x 8 = 16m^2

20 (C) Try each possible answer.

21 (D) Find the number of squares systematically
No. with only 1 small square = 4 x 4 = 16
No. with 4 small squares = 9
No. with 9 small squares = 4
No. with 16 small squares = 1
Total = (16 + 9 + 4 + 1) = 30

22 (B) From any vertex, there are 3 vertices (including itself) to which diagonals can not be drawn. That leaves for a 50-sided polygon, only (50 - 3) = 47 vertices to which diagonals can be drawn from any vertex.

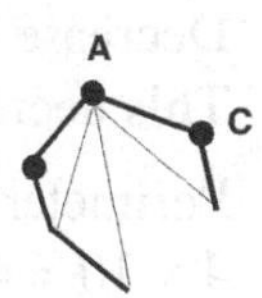

23 (C) Try each possible answer.

24 (D) Both Liz's and Joy's numbers cannot be primes because the first are divisible by 5 and the second by 2.
Fiona's numbers could be 17, 27, 37, 47, 57, 67, 77, 87, 97, of which 17, 37, 47, 67, 97 are prime.
Anne's numbers could be 19, 29, 39, 49, 59, 69, 79, 89, 99; of which 19, 29, 59, 79, 89 are prime.

25 (B) 3 pies and 2 drinks cost \$8.80
2 pies and 3 drinks cost \$7.20
Thus 5 pies and 5 drinks cost (\$8.80 + \$7.20) = \$16
Hence 1 pie and 1 drink costs \$16 ÷5 = \$3.20

PAPER 14

1 (C) We test each given statement to see if it is false.
(2 x 3) x 3 = (2) x 3 = 2; 2 x 3 = 2 and 2 x 1 = 2; 2 x 2 = 0 (Not 1)
Although the given table is a 'multiplication' table, we can use it to do certain divisions. Thus 2 ÷ 3 = 2, since 3 x 2 = 2, just as in ordinary arithmetic we know that say 20 ÷ 5 = 4 since by multiplication 5 x 4 = 20.

2 (A) Vol. of each small cube is $1cm^3$, and thus volume V of solid cube is $125cm^3$.
Length of edge of solid cube = $\sqrt[3]{V} = \sqrt[3]{125}$ = 5cm.

3 (B) The sketch shows the solid cube of 125 small cubes and edge of length 5cm.
The no. of 1cm cubes with just one face painted red is 3 x 3 = 9 for the top and 4 x 3 = 12 for each side face.
This gives a total of 9 + 4 x 12 = 9 + 48 = 57.

4 (B) When 6 dice are thrown simultaneously, the lowest score is 6 x 1 = 6 and the highest score is 6 x 6 = 36. The different totals possible are 6, 7, 8, 9, …, 36. There are (36 - 5) = 31 different totals.

5 (D) Possible different fractions less than 1 are $\frac{1}{2}$, $\frac{1}{3}$, $\frac{1}{4}$, $\frac{1}{5}$, $\frac{2}{3}$, $\frac{2}{5}$, $\frac{3}{4}$, $\frac{3}{5}$, $\frac{4}{5}$ (Note $\frac{2}{4} = \frac{1}{2}$).
There are 9 fractions.

6 (C) Average of these fractions
$= \frac{1}{6}\{\frac{1}{2} + \frac{1}{3} + \frac{1}{4} + \frac{1}{5} + \frac{2}{3} + \frac{2}{5} + \frac{3}{4} + \frac{3}{5} + \frac{4}{5}\} = \frac{1}{6}\{\frac{1}{2} + (\frac{1}{3} + \frac{2}{3}) + (\frac{1}{4} + \frac{3}{4}) + (\frac{1}{5} + \frac{2}{5} + \frac{3}{5} + \frac{4}{5})\}$
$= \frac{1}{6}(\frac{1}{2} + 1 + 1 + 2) = \frac{1}{6} \times 4\frac{1}{2} = \frac{1}{2}$

7 (C) Side of original square = $\sqrt{100}$ = 10cm.
Side of new square = (100% - 20%) of 10cm = 80% of 10cm = $\frac{80}{100} \times 10$ = 8cm.
Thus area of new square = 8^2 = 64cm^2.

8 (B) Decrease in area between original and new square is (100 - 64) = 36cm^2.
This decrease as a percentage of original square = $\frac{36}{100} \times 100\%$ = 36%.

9 (A) Perimeter of new square is 4 x 8 = 32cm and perimeter of original square is 4 x 10 = 40cm.
Required fraction = $\frac{32}{40} = \frac{8}{10} = \frac{4}{5}$.

10 (C) We determine the number of factors of each possible answer.
Now 16 = 1 x 16 = 2 x 8 = 4 x 4; there are 5 factors of 16; namely 1, 2, 4, 8, 16.
Now 36 = 1 x 36 = 2 x 18 = 3 x 12 = 4 x 9 = 6 x 6; there are 9 factors of 36, namely 1, 2, 3, 4, 6, 9, 12, 18, 36.
Now 72 = 1 x 72 = 2 x 36 = 3 x 24 = 4 x 18 = 6 x 12 = 8 x 9; there are 12 factors of 72, namely 1, 2, 3, 4, 6, 8, 9, 12, 18, 24, 36, 72.
Now 128 = 1 x 128 = 2 x 64 = 4 x 32 = 8 x 16; there are 8 factors of 128, namely 1, 2, 4, 8, 16, 32, 64, 128.

11 (A) Counting numbers are 11, 101, 110, 111, 112, 113, 114, 115, 116, 117, 118, 119, 121, 131, 141, 151, 161, 171, 181, 191. There are 20 of these.

12 (D) From 1 to 200 inclusive, there are 200 counting numbers of which from Q11, 20 contain the digit 1 at least twice (i.e. 2 or 3 times).
Thus no. of counting numbers which contain the digit either once or not at all is (200 - 20) = 180.

13 (D) Try each pair of possible answers.

14 (B) Now 80°R is equivalent to 100°C
i.e. 1°R is equivalent to 100/80 = $\frac{5}{4}$°C
Thus 24°R is equivalent to $\frac{5}{4}$ x 24 = 30°C

15 (B) Now 210 = 21 x 10 = (3 x 7) x (2 x 5) = (3 x 5)x (7 x 2) = 15 x 14
The ages of the two teenagers are fifteen and fourteen and their sum is (15 + 14) = 29 years.

16 (A) If the team won 5 out of every 7 games, then they lost 2 games out of each 7 games.
Hence they lost 5 x 2 = 10 games out of 5 x 7 = 35 games.

17 (D) Package price for accommodation only = \$541.64 - \$160 = \$381.64.
Fare for hotel is per night; 8 days and 7 nights means Alan stayed there 7 nights and was there during part of the next day, (but not next night). He is not charged for this day provided he vacates the room at a time stated by the hotel.
Cost of accommodation per night = \$381.64 ÷ 7 = \$54.52

18 (D) The whole numbers to be counted are 1, 2, 3, …, 100.

$$
\begin{aligned}
\text{Let} \quad S &= 1 + 2 + 3 + 4 + \ldots + 99 + 100 \\
\text{Then} \quad S &= 100 + 99 + 98 + 97 + \ldots + 2 + 1, \text{ in reverse} \\
\text{Thus} \quad 2S &= (1 + 100) + (2 + 99) + (3 + 98) + \ldots + (99 + 2) + (100 + 1), \text{ adding} \\
&= 100 \text{ lots of } 101, \text{ i.e. } 100 \times 101 \\
\text{i.e.} \quad S &= \tfrac{1}{2} \times 100 \times 101 = 50 \times 101 = 5050
\end{aligned}
$$

4	3	3

19 (D) No. of colours which can be chosen for first strip is 4; no. of different colours to colour of first strip that can be used for second strip is 3.
Thus, the first 2 strips can be coloured in 4 x 3 = 12 different ways.
No. of colours which can be used for third strip, so it is different to the second strip, is 3. Thus, no of different flags possible = (4 x 3) x 3 = 36.
{For example, if colours are identified as R, B, Y, W then the 12 combinations for first 2 strips are RB, RY, RW; BR, BY, BW; YR, YB, YW; WR, WB, WY.
Each of these 12 combinations must be linked with a third colour not the same as the second colour. Thus RB can be linked with R, Y, W giving RBR, RBY, RBW and so on}.

20 (C) He got back $9.50 for each $4.50 invested,
i.e. $19 for each $9 invested.
This is $1 for each $$\frac{9}{19}$ invested
i.e. $228 for each $$\frac{9}{19}$ x 228 = $108 invested

21 (A) Total distance travelled (170 + 240) = 410km
Time = 3h15min. + 40min. + 4h20min. = 8h15min. = 8¼h.

$$\text{Average speed} = \frac{410}{8\frac{1}{4}} = \frac{410 \times 4}{8\frac{1}{4} \times 4} = \frac{1640}{33} = 49 \text{ rem } 23, \text{ by division}$$

= 50km/h to nearest km/h

22 (A) Using sign < to indicate ' is shorter than' and sign > to indicate 'is taller than', then we are told P<Q, R>S, S<P.
From this we can conclude S<P<Q, i.e. Q>P>S but although we are given R>S, we do not know how tall R is relative to P, Q. We can only say one of the following R>Q>P>S or Q>R>P>S or Q>P>R>S is true. That is, the photographer can arrange the students in the order RQPS, QRPS or QPRS. In only one of these 3 possible arrangements is S 2 students away from R, namely RQPS.

23 (B) Let same time be 1h, so that Mike can ride at 6km/h and Carol at 4km/h.
Mike can cover the 18km in 18 ÷6 = 3h, and Carol can cover it in 18 ÷ 4 = 4½h.
Thus Mike must give Carol 1½h start; in this time Carol can cover 4 x 1½ = 6km.

24 (A) Try each possible answer.
Thus if Tom has say 175 comics, then Deborah has 175 ÷ 5 = 35 comics. Since Deborah has 40% = ⅖ of the no. of comics owned by George, then George has 5/2 of the no. of comics owned by Deborah, i.e. 5/2 x 35 = 87½ - this is obviously incorrect.

25 (D) Total weight of 4 blocks = 4 x 1.5 = 6kg
Obviously the heaviest weight cannot be 6kg for there are 3 other weights to account for.
Also, if the heaviest weight is 1.5kg, the other 3 weights are less than this, and hence this total weight is less than 6kg. Thus, the average of the 4 weights must be less than 1.5kg.
The heaviest weight cannot be 8kg, since the 4 weights together only weigh 6kg.
The heaviest weight could be 3kg, with the other 3 weights together totalling 3kg, to make a total of 6kg.

PAPER 15

1 (D) Write down numbers in a systematic way.
Thus 1 digit nos. are 2, 5, 8
2 digit nos. are 25, 28, 52, 58, 82, 85
3 digit nos. are 258, 285, 528, 582, 825, 852.

2 (C) On 1 June, original length 42mm reduced by 3mm.
On 2 June, original length has been reduced by 2 x 3 = 6mm.
On 3 June, original length has been reduced by 3 x 3 = 9mm.
On 4 June, original length has been reduced by 4 x 3 = 12mm.
....
On 13 June, original length has been reduced by 13 x 3 = 39mm.
Then, only (42 - 39) = 3mm left, and this is burnt on next day 14th June.

3 (A) Certain number is 60 ÷ 4 = 15; correct answer should be 15 ÷ 4 = 3.75

4 (A)

Players say	1__2 3__4 5__6 7__8 9__10 11__12 13__14
After 7 matches, winners say	1_____3 5_____7 9______11 13 (bye)
After 3 matches, winners say	1__________5 9__________13
After 2 matches, winners say	1______________________9
After 1 match, winner say	1

There are (7 + 3 + 2 + 1) = 13 matches needed.

5 (C) Try each answer.
Thus if take $1\frac{1}{3} = \frac{4}{3}$, reciprocal is $\frac{3}{4}$ (since $\frac{4}{3} \times \frac{3}{4} = 1$)
Sum = $\frac{4}{3} + \frac{3}{4} = \frac{16+9}{12} = \frac{25}{12}$. Not correct

6 (D) Sum of 50 numbers is 50 x 38 = 1 900. When the numbers 45, 55 are removed, there are 48 numbers left with sum 1 900 - (45 + 55) = 1 800.
Mean of these numbers = $\frac{1800}{48} = \frac{300}{8} = \frac{75}{2} = 37.5$

7 (B) 4*6 = (4 - 1) x (6 + 1) = 3 x 7 = 21

8 (B) 3*k = (3 - 1) x (k + 1) = 2(k + 1). If 2(k + 1) = 10, k = 4.

9 (B) Digital sum of 1905 is 6 since 1 + 9 + 0 + 5 = 15 and 1 + 5 = 6

10 (B) Now divisors of 2 are 1, 2 ; their sum is 3; 3 ≠ 2 x 2
Also divisors of 6 are 1, 2, 3, 6 ; sum is 12 and 12 = 2 x 6;
this is the smallest perfect number. Try other numbers for practice.

11 (B) From sketch, least number of triangles is 4.

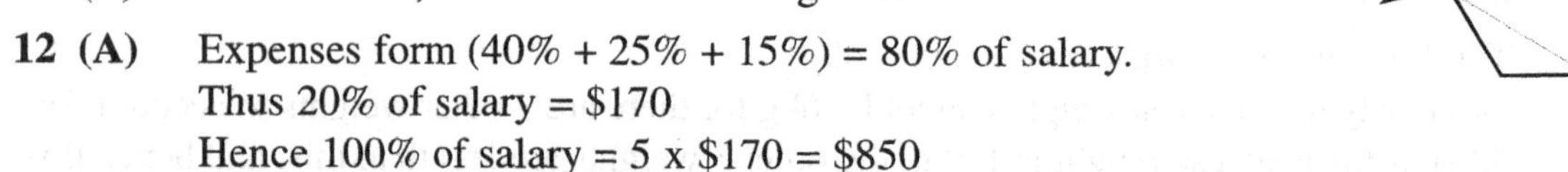

12 (A) Expenses form (40% + 25% + 15%) = 80% of salary.
Thus 20% of salary = $170
Hence 100% of salary = 5 x $170 = $850

13 (D) Now 2.9m = 2.9 x 100 = 290cm

No. of lengths $= \frac{290}{3.8} = \frac{290 \times 10}{3.8 \times 10} = \frac{2900}{38} = \frac{1450}{19}$

= 76 rem 6, i.e. about 76

14 (D) Note 25cm space = ½ of 50cm space.

Plants in each row occupy (1 000cm - 2 x 25cm) = 950cm
No. of 50cm spaces in row = 950 ÷ 50 = 19
No. of plants in each row = 19 + 1 = 20. Check this!
No. of plants in 12 rows = 12 x 20 = 240

15 (D) Length SRQ = 47 - (9 + 16) = 22cm
Length QR = ½ x 22 = 11cm

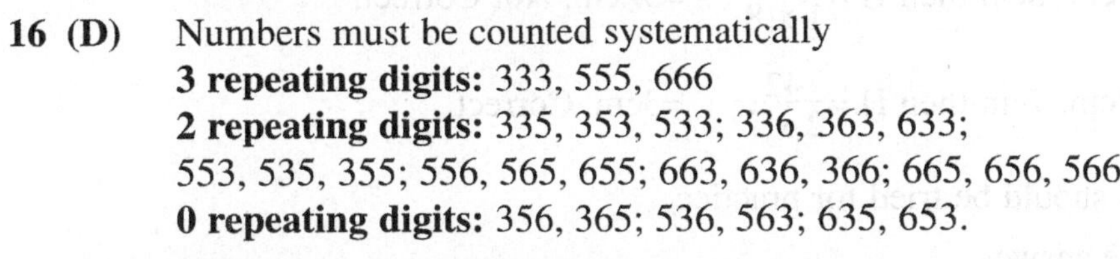

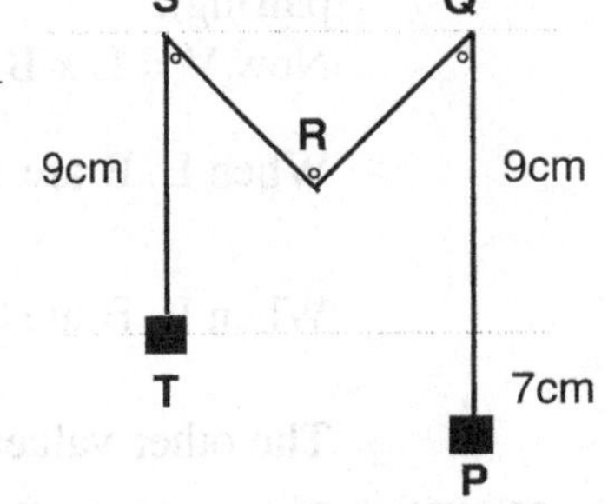

16 (D) Numbers must be counted systematically
3 repeating digits: 333, 555, 666
2 repeating digits: 335, 353, 533; 336, 363, 633;
553, 535, 355; 556, 565, 655; 663, 636, 366; 665, 656, 566
0 repeating digits: 356, 365; 536, 563; 635, 653.

17 (D) Counting must be done carefully.
Squares: consisting of 1 square - no. 10
consisting of 4 squares - no. 2
Rectangles: consisting of 2 squares - no. 11
consisting of 3 squares - no. 6
consisting of 4 squares - no. 3
consisting of 5 squares - no. 1
consisting of 6 squares - no. 1

18 (B) Distance covered = (3 + 2¾ + 6½) x 80m
= 12¼ x 80m, i.e. 980m
To find starting point, try each vertex.
Thus, if start at R; 3 times around in anti-clockwise direction ends again at R; 2¾ times around in clockwise direction ends at Q and 6½ times around in anti-clockwise direction ends at S.

P Q 20m S R

19 (D) $\frac{9!}{5!4!} = \frac{9 \times 8 \times 7 \times 6 \times 5 \times 4 \times 3 \times 2 \times 1}{5 \times 4 \times 3 \times 2 \times 1 \times 4 \times 3 \times 2 \times 1} = \frac{9 \times 8 \times 7 \times 6}{4 \times 3 \times 2 \times 1} = 3 \times 7 \times 6 = 126$

$\frac{8!}{4!4!} = \frac{8 \times 7 \times 6 \times 5 \times 4 \times 3 \times 2 \times 1}{4 \times 3 \times 2 \times 1 \times 4 \times 3 \times 2 \times 1} = \frac{8 \times 7 \times 6 \times 5}{4 \times 3 \times 2 \times 1} = 7 \times 2 \times 5 = 70$

20 (C) Prizes can be won as follows:
Three by same child i.e. AAA, BBB, CCC
Two by one child, one by another i.e. AAB, AAC, BBA, BBC, CCA, CCB
One by each child i.e. ABC
{Note AAB means Ann wins 2 prizes and Betty wins one; Charles does not win a prize, and so on}

21 (D) Since the rectangular prism is made of 42 cubes of side 1cm, its volume V is $42cm^3$ and the measurements are whole numbers of cm.

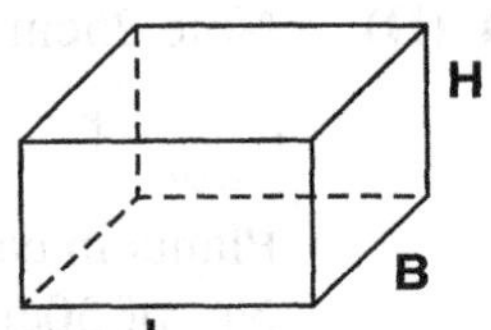

If the base has perimeter 18cm, then L + B = 9 cm.
The possible values of L, B are 1cm, 8cm or 2cm, 7cm or 3cm, 6cm or 4cm, 5cm. (These can be reversed in each pairing).
Now V = L x B x H, i.e. H = $\frac{V}{L \times B}$
When L, B are 1cm, 8cm then H = $\frac{42}{1 \times 8}$ =5¼cm. Not Correct.

When L, B are 2cm, 7cm then H = $\frac{42}{2 \times 7}$ =3cm. Correct.

The other values should be tried for practice.

22 (D) Try each possible answer.
Thus if tank held 50L originally, after each month for 5 months and addition of the extra litre, the tank would hold in turn 26L, 14L, 8L, 5L 3½L. This is incorrect.

23 (D) Increase in surface area is equal to 8 faces each 7cm x 6cm, i.e. 8 x 42 = $336cm^2$. (Taking the blocks from the left, these extra faces are one from block 1, two from each of the blocks 2, 3, 4 and one from block 5).

24 (B) 1.44L = 1.44 x 1 000mL = 1440mL = $1440cm^3$.
Vol. of water in tank from jug = $1440cm^3$.
Depth of water in tank = $\frac{1440}{18 \times 10}$ = 8cm

25 (A) No.s 1, 2, 3, 4, 5, 6, 7, 8, 9, 10 can be written as 1, 2, 3, 2^2, 5, 2 x 3, 7, 2^3, 3^2, 2 x 5.
The smallest number then can be divided by each of these is $2^3 \times 3^2 \times 5 \times 7$ = 8 x 9 x 5 x 7 = 2 520
When 2520 is divided by 11, the remainder is 1.

11) 2520
229 rem 1

PAPER 16

1 (B) Numbers less than 60 which are
multiples of 3 are 3, 6, 9, 12, 15, …, 57
even multiples of 3 are 6, 12, 18, 24, …, 54
odd multiples of 3 are 3, 9, 15, 21, 27, 33, 39, 45, 51, 57.
There are 10 of these.

2 (B) Area of square = area of rectangle = 24 x 6 = $144cm^2$.
Side of square = $\sqrt{144}$ = 12cm.

3 (C) If we denote the symbol > to mean 'can swim faster than', then we are told <u>R>S</u>, <u>P>Q</u>, <u>S>Q</u>, <u>R>P</u>.
From this, we see from the underlined statements that R>S>Q and R>P>Q.
Thus, R is faster than S, Q and P. Hence R should win the race.

4 (D) Let speed of car X be say 60km/h, then distance travelled by it in ½h is ½ x60 = 30km
Car Y is travelling at ½ x 60 = 30km/h and in 1h, travels a distance of 30 x 1 = 30km.
Thus cars X, Y travel the same distance.

5 (C)

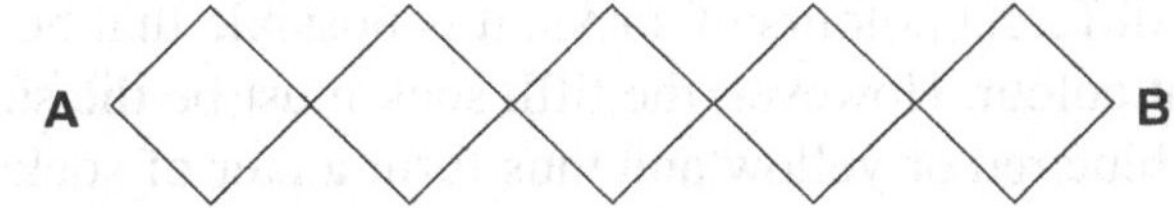

Largest perimeter occurs when squares touch as shown. If we work on lower side of figure from A to B, the distance is 10 x 1 = 10cm, and then return from B to A on upper side of figure. This perimeter is 20cm.

6 (B) If the girl has say 40 - 20c coins, then she has 20 - 5c coins, a total of (40 + 20) = 60 coins.
If she changed the 5c coins she would receive 10 - 10c coins. She then has (40 + 10) = 50 coins. Compared to the original number, she has $\frac{50}{60} = \frac{5}{6}$ as many now.
That is, she has five-sixths as many coins as before.

7 (A) Try each answer in turn to see if t - u = u - t.
If t = u, i.e. say t = 5, u = 5 we have 5 -5 = 5 -5, which is true.
If t = 0 and u ≠ 0, (say u = 5) then 0 - 5 ≠ 5 - 0
If u = 0 and t ≠ 0, (say t = 5) then 5 - 0 ≠ 0 - 5
If t, u can take all possible values say t = 7, u = 4 then 7 -4 ≠ 4 - 7
For no values of t or u is incorrect since t - u = u - t when t = 7, u = 7 say.

8 (C) bottle + jam weigh 541g
bottle + ½jam weigh 315g
Thus ½ jam weighs (541 - 315) = 226g
i.e. full jam weighs 226 x 2 = 452g
Hence bottle alone weighs (541 - 452) = 89g.

9 (A) In a mirror, the hands of the clock are reflected in the vertical line through 12 and 6. The following sketches show the possible times given and the corresponding clock faces in the mirror.

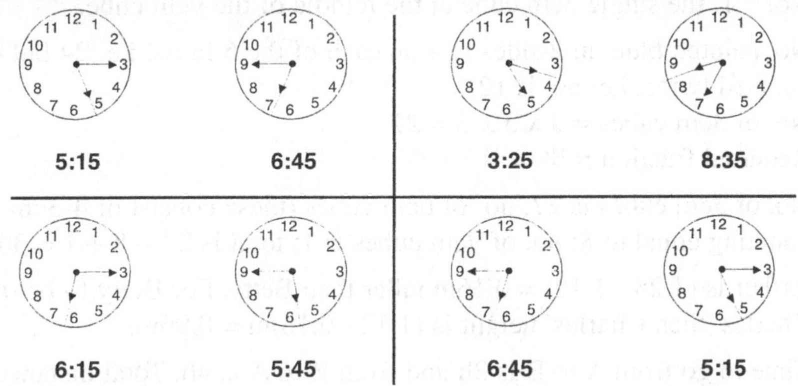

10 (C) No. of spaces 10m apart along the 150 m fence is 150 ÷ 10 = 15. Since there is a post at each end of this fence the number of posts is 15 + 1 = 16.

120m

150m

11 (B) The number of posts in the 120m fence is 120 ÷ 10 + 1 = 13
No. of posts around area = (2 x 16 + 2 x 13) - 4 = 54, noting the 4 corner posts have been counted twice

12 (D) Since these are 4 different colours of socks, it is possible that he can take out 4 socks each of a different colour. However, the fifth sock must be the same colour as one of the three colours blue,red or yellow and thus form a pair of socks with the same colour. The greatest number required is 5.

13 (B) During 1h, the tap could fill 5 similar bathtups.
During 1h also, 4 similar bathtubs could be emptied.
Hence together (filling and emptying) 1 bathtub would be filled in 1h,
i.e. in 60 minutes.

14 (B) If n is a whole number then
3 x n can be odd or even; e.g. 3 x 7 = 21, 3 x 4 = 12
2 x n + 1 must be odd; e.g. 2 x 7 + 1 = 15, 2 x 4 + 1 = 9
$\mathbf{n^2}$ can be odd or even; e.g. $7^2 = 49$, $4^2 = 16$
n x n x n can be odd or even; e.g. 3 x 3 x 3 = 27, 2 x 2 x 2 = 8

15 (C) Bell 1 rings after 10, 20, 30, 40, 50, <u>60</u>, 70, ... minutes
Bell 2 rings after 12, 24, 36, 48, <u>60</u>, 72, ... minutes
The bells ring together again after 60 minutes (Note 60 is the L.C.M. of 10 and 12).

16 (B) Now 12.3 x 4.8 + 16.4 ≑ 12 x 5 + 16 = 76
Also 5.9 x 3.6 ≑ 6 x 3.5 = 21
Fraction ≑ $^{76}/_{21}$ ≑ $^{75}/_{20}$ = 3.75 which is nearest to 4.

17 (B) Try a number, e.g. 46 becomes 465; the 465 is 10 times 46 plus 5.
Try another, e.g. 67 becomes 675; the 675 is 10 times 67 plus 5

18 (C) Minimun no. of cuts needed is (2 + 2) = 4 vertically and 2 horizontally; total 6.

19 (B) No. = 4 corner cubes at top and 4 corner cubes at base; total 8

20 (B) No. = middle one on each 6 faces; total 6

21 (B) No. = 1, the single 3cm cube at the middle of the 9cm cube.

22 (C) No. painted blue on 2 sides is 4 on each of the 6 faces, i.e. 24 but each cube has been counted twice; i.e. no. is 12.
No. of 3cm cubes = 3 x 3 x 3 = 27
Required fraction = $^{12}/_{27}$ = $^{4}/_{9}$

23 (B) No. of 3cm cubes is 27; no. of 6cm cubes (these consist of 4-3cm cubes) is by careful counting equal to 8; no. of 9cm cubes is 1; total is 27 + 8 + 1 = 36

24 (B) Arthur is (1.28 - 1.12) = 0.16m taller than Betty. For Betty to be 0.16m taller than Charles, then Charles' height is (1.12 - 0.16)m = 0.96m.

25 (C) Time to go from A to B is 3h and from B to A is 4h. Total distance covered is 480km and total time for the trip is 7h. Average speed is 480/7 = 68$^{4}/_{7}$kmh.

PAPER 17

1 (D) Try each possible answer in turn.
Thus, if no. changing is 6, then (20 + 6) = 26 students play football out of class of 30.
Percentage = $^{26}/_{30}$ x 100% = $^{260}/_{3}$% = 86$^{2}/_{3}$%. This is not correct answer.

2 (B) Try each answer.
Thus if no. ends in 2 or 3, the squares end in 4 or 9 - not correct

3 (A) Dick gets \$30, then Harry gets \$30 + \$15 = \$45, whilst Tom gets 2 x \$30 = \$60.
Sum is \$(30 + 45 + 60) = \$135.

4 (D) Work as percentages (or decimals) $\frac{3}{4} = 75\%$, $\frac{3}{5} = 60\%$.
Now $\frac{12}{25} = \frac{12}{25} \times 100\% = 48\%$, $\frac{17}{20} = 85\%$, $\frac{8}{15} = \frac{8}{15} \times 100\% = \frac{160}{3}\% = 53\frac{1}{3}\%$; $\frac{2}{3} = 66\frac{2}{3}\%$.

5 (D) There are 2 options for light 1, it can be on or off. This applies for each light.
The no. of different signals = 2 x 2 x 2 x 2 x 2 = 32, but this includes the signal when all 5 lights are off. Thus no. of signals is 32 -1 = 31.
{One of the 31 signals is given in question - it shows lights as off, off, on, off, on}

6 (C) Increase of 10% in \$50 skirt makes it (\$50 + \$5) = \$55.
Decrease of 10% in \$55 makes price (\$55 - \$5.50) = \$49.50

7 (A) Sides of cubes are $\sqrt[3]{1}$, $\sqrt[3]{8}$ = 1, 2 cm respectively.

8 (C)

9 (C) Area unshaded triangle = ½base x perp. height, i.e. ½ x 12 x 8m²
But area rectangle = 12 x 8m²
Thus area of unshaded triangle is half area of rectangle, and hence shaded area is also half area of rectangle, i.e. ½ x 96 = 48m².

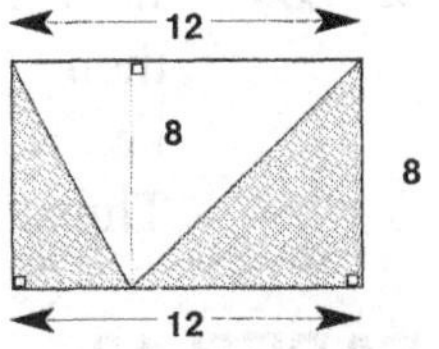

10 (C) **11 (A)**

12 (A) $\frac{2}{3}$ of my pocket money = \$4
i.e. $\frac{1}{3}$ of my pocket money = \$2
$\frac{3}{3}$ i.e. my pocket money = \$6

13 (D) $\frac{3}{5}$ of my salary was \$17 640
$\frac{1}{5}$ of my salary was \$17 640 ÷ 3 = \$5 880
$\frac{5}{5}$ of my salary was \$5 880 x 5 = \$29 400.

14 (B) There are 26 possible letters for the missing letter and 10 possible digits for the missing number. Thus, there are 26 x 10 = 260 number plates, and hence 260 possible cars. □XA ○37

15 (B) The 2 numbers add to 42, and their difference is 8. Now the equal numbers 21, 21 add to 42, but their difference is 0. Add 4 to the first number and deduct 4 from the second number; this gives numbers 25, 17 which differ by 8.
Their product is 25 x 17 = 425.

16 (B) Try each possible answer in turn.
Thus if number is 4, then ½(4 + 11) = 7½ whilst 4 + 3 = 7; this is not correct.

17 (B)

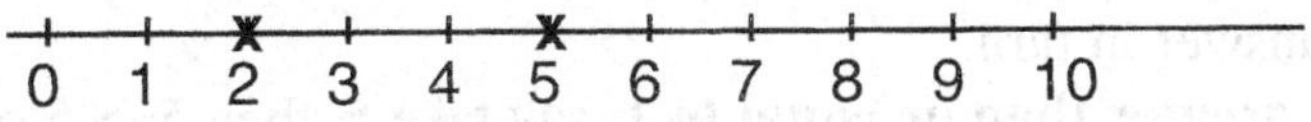

Try each possible answer in turn.
Thus 3 is 1 unit from 2 and 2 units from 5; since 1 ≠ 2 x 2; then 3 is not correct.
Similarly 6 is incorrect.

18 (A) Follow the instructions carefully.

19 (D) Try each possible answer in turn. (KY) x (KY) = ZLK
If K = 1, then (1Y) x (1Y) = ZL1 and hence Y x Y gives a 1.
This is true if Y = 1 or 9. If Y = 1, then number is 11, and product 11 x 11 = 121 whilst if Y = 9, number is 19 and product 19 x 19 = 361. Both of these are 3 digit numbers are of the form ZLK. Thus K = 1 is a possible value for K.
Now if K = 9, then (9Y) x (9Y) = ZL9 and hence Y xY gives a 9. this is true for Y = 3 or 7. If Y = 3, number is 93 and product 93 x 93 = 8 649 whilst if Y = 7, number is 97 and product 97 x 97 = 9 409. Both of these answers 8 649, 9 409 have 4 not 3, digits, and thus K = 9 is incorrect.

20 (B) Shapes (A), (C), (D) can be obtained by rotating shape K in the plane of the paper. However, to obtain shape (B), then shape K must be flipped over.

21 (C) $\frac{1}{3} \times \frac{1}{3} \div \frac{1}{3} = \frac{1}{3}$ (cancelling $\frac{1}{3}$'s)

22 (B) 24kg + $\frac{1}{2}$ weight of dog = weight of dog
i.e. $\frac{1}{2}$ weight of dog is 24kg and weight is 48kg.

23 (A) 15% of original price of can = \$12
i.e. 5% of original price of can = \$12 ÷ 3 = \$4
Thus 100% i.e. original price of can = 20 x \$4 = \$80

24 (D) 13 000 + 1 300 + 13 = 14 313. Digit is 4.

25 (D) If 9 x (98 7□5) + 3 = 888 888
then 9 x (98 7□5) = 888 885
i.e. 98 7□5 = 888 885 ÷ 9 = 98 765
Thus □ = 6.

PAPER 18

1 (B) A cube has 12 edges, such as AB.
Each of these 12 edges is parallel to 3 edges,
e.g. AB || DC, AB || EF, AB || HG.
This gives 12 x 3 = 36 pairs of parallel edges. But each of these pairs has been counted twice, e.g. AB || DC and DC || AB.
This leaves 36 ÷ 2 = 18 pairs of parallel edges.

2 (B) Try different numbers greater than 4 for □.
Thus, if □ = 5 say, the respective values of the expressions given are $\frac{9}{5}$, 2, $\frac{5}{9}$, $\frac{5}{8}$; the largest is 2.
Similarly, try □ = 8½ say, their respective values are $\frac{16}{5}$, $\frac{17}{5}$, $\frac{5}{16}$, $\frac{5}{15}$; the largest is $\frac{17}{5}$.
That is, the greatest number is $\frac{2 \times \square}{5}$.

3 (C) If exactly 35% of a group agreed with a proposal, then 35% of the group is a whole number of persons.
Now 35% = $\frac{7}{20}$; i.e. $\frac{7}{20}$ of group is a whole number and hence the number of persons in the group must be one of 20, 40, 60, 80, …. The smallest number in the group is 20.

4 (B) Try each posssible answer in turn.
Thus, for numbers greater than or equal to 1, say take 5, then $5^2 > 5$ or take 1, then $1^2 = 1$. These squares are not less than the number.
Try numbers between 0 and 1, say take $\frac{1}{3}$, then $(\frac{1}{3})^2 = \frac{1}{9}$ which is less than $\frac{1}{3}$ or take $\frac{2}{5}$, then $(\frac{2}{5})^2 = \frac{4}{25}$ which is less than $\frac{2}{5}$ ($\frac{10}{25}$). These squares are less than the number.

5 (C) Try some possible answers.
Thus 5, 7, 9 are consecutive odd numbers whose sum is 21. This sum is odd and divisible by 3.
Try say 41, 43, 45; the sum is 129. This sum is odd and divisible by 3.

6 (D) If a, b are counting numbers and a + b is odd, then one of a or b is odd and the other is even; for example 5 + 8 = 13. However a x b is even not odd, thus here 5 x 8 = 40. That is, we cannot find counting numbers a, b such that a + b is odd and a x b is odd as well.
If instead a x b is odd, then both a, b are odd, e.g. 5 x 9 = 45 but a + b is even (not odd), thus here 5 + 9 =14. Thus, we cannot find counting numbers a, b so that a x b is odd and a + b is odd as well.

7 (D) Test each possible answer to see if it could give a prime number.
Thus R + R = 2 x R, which is divisible by 2 or R, and hence not prime.
Also 3 x P X Q is divisible by 3 or P or Q, and thus not prime.
For P x Q + Q consider specific values for P, Q; say P = 8, Q = 5
then P x Q + Q = 8 x 5 + 5 = 45, which is not prime.
Now P + 1 could be prime; e.g. if P = 24, then P + 1 = 25 which is not prime but if P = 12, then P + 1 = 13 which is prime.

8 (D) Consider easy values for the length and breadth of the original rectangle, say 20cm, 15cm.
Area of original rectangle is 20 x 15 = 300cm^2.
After 10% increase, new length = 20 + 2 = 22cm
After 20% increase, new breadth = 15 + 3 = 18cm
Area of new rectangle = 22 x 18 = 396cm^2
Increase in area of original area = (396 - 300) = 96cm^2
This percentage increase in area = $^{96}/_{300}$ x 100% = $^{96}/_{3}$% = 32%

9 (B) Each child gives 5 presents and thus for 6 children there are 6 x 5 = 30 presents.

10 (A) Angle at centre for shaded parts = 360° - 2 x 30° = 300°.
Fraction of circle shaded = $^{300}/_{360}$ = $^{5}/_{6}$

11 (D) The usual mistake is to say the bottle costs 12c and the cork 1c. This gives a total of 13c, but the bottle costs 11c, not 12c, more than the cork.
To get the correct answer, we simply add ½c to the cost of the bottle making it 12½c and deduct ½c from the cost of the cork making it ½c.

12 (C) Total age of 10 persons is 10 x 14 = 140 years, and of the remaining 8 persons is 8 x 15 = 120 years. The two persons have ages totalling (140 - 120) = 20 years. and hence since they are the same age, each is of age 10 years.

13 (B) Average speed to travel 720km in 9h is $^{720}/_{9}$ = 80km/h.
Average speed to travel 720km in 8h is $^{720}/_{8}$ = 90km/h.
Difference in speeds is 10km/h.

14 (B) Can +8L oil weigh 42kg
Can +3L oil weigh 27kg
Thus 5L oil weigh (42 - 27) = 15kg
i.e. 8L oil weigh 3 x 8 = 24kg
That is, can weighs (42 - 24) = 18kg.

15 (A) If 24 men can construct a kit home in 18 days
then 1 man can construct it in 18 x 24 days
and 27 men can construct it in $\frac{18 \times 24}{27}$ = 16 days

16 (B) Consider specific cases and test results.
Thus take first square with perimeter 4cm and second square with perimeter 8cm.
Sides of squares are respectively 1cm, 2cm.
Area of first square is 1 x 1 = 1cm^2 and of second square is 2 x 2 = 4cm^2. That is, area of larger square is four times as large as the smaller square.

17 (C) After spending $\frac{2}{5}$ of his money, he had $\frac{3}{5}$ of his money left.
Now $\frac{2}{5}$ of this remainder = $\frac{2}{5} \times \frac{3}{5} = \frac{6}{25}$ of his money.
Here we know $\frac{6}{25}$ of his money is \$1.20
i.e. $\frac{1}{25}$ of his money is \$1.20 ÷ 6 = 20c
Thus $\frac{25}{25}$ i.e. his money is 25 x 20c = \$5.00

18 (D) Now $\frac{2}{3} - \frac{1}{2} = \frac{1}{6}$
Thus $\frac{1}{6}$ of full bottle is 450mL
$\frac{6}{6}$ i.e. full bottle, holds 6 x 450 = 2 700mL

19 (C) No. of choices of entrees is 4, of main courses is 5 and of desserts is 3.
No. of different 3 course meals possible is 4 x 5 x 3 = 60.

20 (B) The weights in the boxes in the first diagram given, balance one another,
i.e. [X] + [X] + [X] + [5] kg balance [X] + [17] kg.
If we take one of the boxes indicated [X] kg from each balance,
we have [X] + [X] + [5] kg balancing [17] kg. At sight, we can see that [X] = [6] kg.
Considering the second diagram given, the [?] kg balances
[X] + [X] + [X] + [X] + [10] kg; hence [?] is equal to [6] + [6] + [6] + [6] + [10],
i.e. [34] kg.
{Note. We can shorten this reasoning by using Algebra.
From the first balance, using kg,

x + x + x + 5 = x + 17,	i.e.	3x + 5 = x + 17
Subtract x from both sides,	i.e.	2x + 5 = 17
Subtract 5 from both sides,	i.e.	2x = 12
Divide both sides by 2,	i.e.	x = 6

From the second balance, ? = 4x + 10, i.e. 4 x 6 + 10
= 34}

21 (A) In 1h car travels 54km
i.e. in 60min. car travels 54 000m
Thus, in 1min. car travels 54 000 ÷ 60 = 900m
i.e. in 60s car travels 900m
Thus, in 1s car travels 900 ÷ 60 = 15m
Hence in 12s car travels 15 x 12 = 180m

22 (D) Try each possible answer in turn
Thus, if 15 hens and hence (37 - 15) =22 sheep, there are
(15 x 2 + 22 x 4) = 30 + 88 = 118 legs. Not correct.

23 (D) Each step brings them (70 + 80) = 150cm closer.
After 224 steps, they will have covered 150 x 224cm i.e. $\frac{150 \times 224}{100}$ m = 336m.

24 (C) If an amount \$A is increased by 20%, then the new amount is (100% + 20%) = 120% of \$A.
Here the new price of the pen is 120% of the original price,
i.e. 120% of original price is \$5.82 by data
then 20% of original price is \$5.82 ÷ 6 = 97c
and 100% i.e. original price, is 97c x 5 = \$4.85

25 (D) Try each possible answer in turn.
Thus, if original number is 15, after increase by 25% its value is (15 + 3.75) = 18.75.
If this is decreased by 25%, its value is (18.75 - 4.6875) ≠ 15. Not correct.

PAPER 19

1 (D) If 6 x □ + 4 = 12, then 6 x □ = 8 and □ = $\frac{8}{6}$ = $1\frac{1}{3}$.
Then 8 x □ + 5 = 8 x $1\frac{1}{3}$ + 5 = $10\frac{2}{3}$ + 5 = $15\frac{2}{3}$.

2 (C) The 3 in 231 denotes 30 and the 3 in 491.3 denotes $\frac{3}{10}$. Note 30 = 100 x $\frac{3}{10}$.

3 (D) Original distance = 66 x $\frac{1}{3}$ = 22km; new distance = 44km.
Speed now = $\frac{44}{\frac{1}{2}} = \frac{44 \times 2}{\frac{1}{2} \times 2} = \frac{88}{1}$ = 88km/h

4 (D) The differences in terms of sequence 341, 213, 149, 117, 101, … are respectively 128, 64, 32, 16. The next term is 101 - 8 = 93.

5 (C) Fraction of class playing cricket = $\frac{3}{5}$ + $\frac{1}{2}$ of $\frac{2}{5}$ = $\frac{3}{5}$ + $\frac{1}{5}$ = $\frac{4}{5}$.
Thus $\frac{4}{5}$ of class = 12 students
i.e. whole class = $\frac{12}{4}$ x 5 = 15 students

6 (A) Thickness of folded paper after 1 fold = 2 x 0.3 = 0.6mm
" " 2 folds = 2 x 0.6 = 1.2mm
" " 3 folds = 2 x 1.2 = 2.4mm
Thus thickness after 4, 5, 6, 7 folds is respectively 4.8mm, 9.6mm, 19.2mm, 38.4mm = 3.84cm.

7 (A) Surface area of each card = 2 x 4.5 x 8.5 = 76.5cm^2
Surface area of 52 cards = 52 x 76.5 = 3 978cm^2

8 (D) Bathroom scales show (100% + 10%) = 110% of my weight.
Here 110% of my weight is 92.4kg
i.e. 10% of my weight is 92.4 ÷ 11 = 8.4kg
Thus 100% i.e. my weight is 8.4 x 10 = 84kg

9 (D)

10 (A) The cube has 6 faces, each of surface area 216 ÷ 6 = 36m^2
Length of each edge of cube= $\sqrt{36}$ = 6m
Volume of cube = 6^3 = 6 x 6 x 6 = 216m^3

11 (B) Children receive respectively \$9, \$6, \$2 - a total of \$17.
The extra \$1 is added to the \$2 so that child receives \$3.

12 (D) No. of codewords each of 1 letter is 3, of 2 letters is 6 and of 3 letters is 6. Total 15.

13 (C) Try each possible answer in turn.
Thus if there are 8 - 20c coins totalling \$1.60, then only (\$1.75 - \$1.60) = 15c remains which can only be from 1 - 10c and 1 - 5c coin. Not correct.

14 (D) Try each possible answer in turn.
If number is x, then $x^2 - x$ may be 30. This is so when $x = 6$,
for then $6^2 - 6 = 36 - 6 = 30$.

15 (D) Fractions possible are $\frac{3}{2}$, $\frac{4}{2}=2$, $\frac{5}{2}$, $\frac{6}{2}=3$, $\frac{4}{3}$, $\frac{5}{3}$, $\frac{5}{4}$, $\frac{6}{5}$.

16 (D) Try each possible result in turn.
Thus, if the sum of the 3 consecutive odd numbers is 345, then the middle of these is $345 \div 3 = 115$. The numbers are 113, 115, 117 (sum is 345). Not answer to question.

17 (B) $1\text{ha} = 10\,000\text{m}^2 = 100\text{m} \times 100\text{m} = 10\,000\text{cm} \times 10\,000\text{cm}$
If 1 cm rain fell on this hectare,
then volume of water on it $= 10\,000 \times 10\,000 \times 1 = 100\,000\,000\text{cm}^3$
$= 100\,000\,000\text{mL}$
Weight of this water $= 100\,000\,000\text{g}$, since 1mL weighs 1g
$= 100\,000\text{kg}$ (1 000g = 1kg)
$= 100$ tonnes (1t = 1 000kg)

18 (A) If breadth is Bcm, then length is $1\frac{1}{2} \times$ Bcm, and area is $1\frac{1}{2} \times B \times B\text{cm}^2$.
Here $1\frac{1}{2} \times B \times B = 96$ and thus $B \times B = \frac{96}{1\frac{1}{2}} = \frac{192}{3} = 64$

Hence $B = 8$ and then $L = 1\frac{1}{2} \times 8 = 12$
That is, the rectangle has length 12cm, breadth 8cm and thus perimeter 40cm.

19 (C) $\frac{5! + 6!}{4!} = \frac{5!}{4!} + \frac{6!}{4!} = \frac{5 \times 4 \times 3 \times 2 \times 1}{4 \times 3 \times 2 \times 1} + \frac{6 \times 5 \times 4 \times 3 \times 2 \times 1}{4 \times 3 \times 2 \times 1}$

$= 5 + 6 \times 5$, i.e. 35

20 (C) Area $\Delta TSU = \frac{1}{2} \times 2 \times 1 = 1\text{cm}^2$
Area rectangle $PQRS = 2 \times 3 = 6\text{cm}^2$
Shaded area $= (6 - 1) = 5\text{cm}^2$
Required fraction $= \frac{5}{6}$

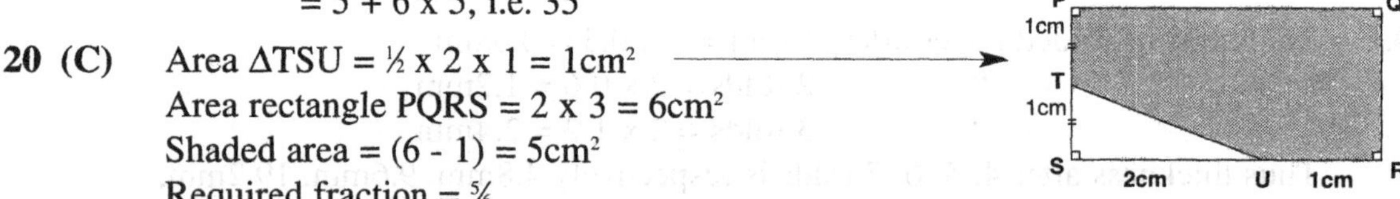

21 (A) The first few primes are 2, 3, 5, 7, 11, 13, 17, 19, 23, 29, ...
Pairs, which sum to 50, which comprise 1 or 2 primes are as follows:
2 + 48, <u>3 +47</u>, 5 + 45, <u>7 + 43</u>, 9 + 41, 11 + 39, <u>13 + 37</u>, 17 + 33, <u>19 + 31</u>, 21 + 29, 23 + 27. (Other pairs repeat these). Required pairs are underlined.

22 (B) If $a*b = a^2 - a \div b$, then $6 * 12 = 6^2 - 6 \div 12$, i.e. $36 - \frac{1}{2} = 35\frac{1}{2}$.

23 (A) The year 2000 is a leap year, and February has 29 days. From 25 February, there are 5 days in February. March has 31 days, April 30 days, May to 10 May is 10 days.
Total no. of days is (5 + 31 + 30 + 10) = 76

24 (C) $1\text{m} = 100\text{cm}$ and thus $1\text{m}^2 = 100 \times 100 = 10\,000\text{cm}^2$
The cube has 6 faces with area of each face 1m^2.
This cube has surface area $6\text{m}^2 = 6 \times 10\,000\text{cm}^2 = 60\,000\text{m}^2$

25 (D) Side of square tray $= \sqrt{400} = 20\text{cm}$
There are 4 coins per side, and hence diameter of each coin is $20 \div 4 = 5\text{cm}$.
Radius of each coin is 2.5cm.

PAPER 20

1 (D) 25% of 40% = $\frac{1}{4}$ of 40% = 10%

2 (D) Fractions are $\frac{6}{25} = \frac{24}{100}$, $\frac{3}{10} = \frac{30}{100}$, $\frac{1}{4} = \frac{25}{100}$, $\frac{1}{5} = \frac{20}{100}$

3 (D) Differences between successive terms of sequence 11, 15, 24, 40, 65 are 4, 9, 16, 25. Next term is 65 + 36 = 101.

4 (A) No. of balls in box is one of 21, 26, 31, 36. Only 26 leaves 2 when divided by 3.

5 (C) Vol. of cube = $6^3 = 216\text{cm}^3$.
Hole has square end of side 4cm, but length of hole is 6cm.
Vol. of hole = 4 x 4 x 6 = 96cm^3
Remaining volume = (216 - 96) = 120cm^3

6 (C) It is best to sketch each possible answer.

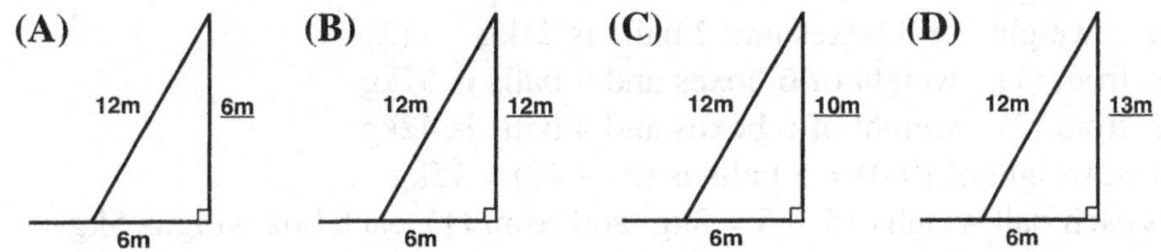

In (A), the 12m ladder should be less than not equal to the sum of lengths 6m, 6m.
In (B), the largest side of a right-angled should be opposite the right-angle; here we have two 12m lengths, one opposite the right-angle.
In (D), the side opposite the right-angle is not the longest side.
The best approximation is shown in (C).

7 (C) Try each possible answer in turn.
Thus if Saul has \$24, Barry has \$24 + \$7 = \$31 and Alla has \$24 + \$19 = \$43.
Their sum is not \$32.

8 (A) Height of man + height of one son is 180cm
Height of man + heights of two sons is 252cm
Thus height of a son is (252 - 180) = 72cm
Hence height of man = (180 - 72) = 108cm
Required answer = (2 x 108 + 72) = 288cm

9 (C) 'Undo' the process. Multiply 40° by 9, add 160 and divide by 5.
This gives (40 x 9 + 160) ÷ 5 = 520 ÷ 5 = 104°F

10 (A) Consider 2 specific numbers, say 4387 and 7384. The latter is larger and their difference is (7384 - 4387) = 2997.

11 (A) Divide the shaded part into rectangles. ⟶
There are many ways to do this; one is shown.
We can easily obtain the further dimensions shown.
Area A = 2 x 1 = 2m^2; Area B = 3 x 2 = 6m^2
Area C = 3 x 1 = 3m^2; Area D = 1 x 3 = 3m^2.
Total area is 14m^2.

12 (A) 64 - 10 = 54 students play cricket or tennis or both.
If we add the number of students 33 who play cricket to the number 28 who play tennis we are counting those students, say x of them, who play both games twice.
This result is 54. i.e. 33 + 28 - x = 54, i.e. 61 - x = 54 and hence x = 7.

13 (B) Try some fractions less than than 1 as required.
Say $\frac{4}{5}$. If we add say 3 to numerator and denominator we obtain $\frac{7}{8}$.
Now $\frac{4}{5}$ = 0.8 and $\frac{7}{8}$ = 0.875; $\frac{7}{8} > \frac{4}{5}$.
Try $\frac{2}{3}$, then adding 7 as above, we obtain $\frac{9}{10}$. Now $\frac{2}{3}$ = 0.666... and $\frac{9}{10}$ = 0.9; $\frac{9}{10} > \frac{2}{3}$.

14 (C) Total percentage for class P is 20 x 60 = 1 200%
Total percentage for class Q is 28 x 70% = 1 960%
Sum of percentages for (20 + 28) = 48 students is (1 200 + 1 960) = 3 160%
Percentage average for 2 classes = $\frac{3160}{48} = \frac{395}{6} = 65\frac{5}{6}$%, which gives 66% to the nearest %.

15 (D) Weight of 2 boxes and 3 balls is 19kg - (1)
Also weight of 3 boxes and 2 balls is 21kg - (2)
now, from (1), weight of 6 boxes and 9 balls is 57kg
and, from (2), weight of 6 boxes and 4 balls is 42kg
That is, weight of (9-4) = 5 balls is (57 - 42) = 15kg
Thus each ball weighs 15 ÷ 5 = 3kg and from (1), each box weighs 5kg
Hence weight of 3 boxes and 4 balls is (15 + 12) = 27kg.

16 (D) We use the notation here > to mean 'is taller than' and < to mean 'is shorter than'.
We are given P > R, Q > S, P < Q, R< S.
That is $\underline{R < P}$, $\underline{S < Q}$, $\underline{P < Q}$, $\underline{\underline{R < S}}$
Thus, from underlined items R< P < Q and R<S<Q. Although we can see that R is the shortest and Q is the tallest boy, we cannot determine who is the shorter (or taller) of P and S. That is, there is not enough information to arrange P, Q, R, S in order of height.

17 (D) Water flows into the tank at 3L/15s, i.e. at 6L per 30s, whilst water flows out at 3L/10s, i.e. at 9L per 30s. Thus every 30s, the tank loses 3L.
Since the tank is half full, it has 24L initially.
To empty the tank, takes 24 ÷ 3 = 8 lots of 30s, i.e. 240s = 4min.

18 (B) From the addition table 3 + 2 = 0 and from the multiplication table 3 x 2 = 1.
Thus 3 + 2 + 3 x 2 = 0 + 1 = 1

19 (D) 3^2 = 3 x 3 = 4 and 2 x (2 + 4 x 3) = 2 x (2 + 2) = 2 x 4 = 3
Thus 3^2 + 2 x (2 + 4 x 3) = 4 + 3 = 2

20 (B) 2^3 = (2 x 2) x 2 = 4 x 2 = 3

21 (B) The square root of 4, i.e. $\sqrt{4}$, is the number(s) which when squared gives 4. From the multiplication table 2^2 = 2 x 2 = 4 and 3^2 = 3 x 3 = 4. Thus $\sqrt{4}$ = 2 or 3.

22 (D) If 2 - 3 = n (say), then 3 + n = 2. From the addition table 3 + 4 = 2, thus n = 4, i.e. 2 -3 = 4.

23 (D) If 2 ÷ 3 = Q, then 3 x Q = 2. From the multiplication table, 3 x 4 = 2.
Thus Q = 4, i.e. 2 ÷ 3 = 4.

24 (B) Now 4 + 1 = 0, and thus K = 1

25 (D) Since 2 + 2 = 4, then if 3 x P + 2 = 4, it follows that 3 x P = 2, whence P =4.

Coroneos Publications – Titles by Topic
BASIC SKILLS SERIES

ISBN	Item	Title	Author
Basic Skills Maths Tests			
9781862941076	42	Further Maths Tests for Selective Schools Scholarship Exams Yrs 5–8	Coroneos,An,Smith
9781862940697	49	Maths Tests for Selective Schools Scholarship Exams Yrs 5–8	Coroneos,An,Smith
Basic Skills Tests			
9781862941243	112	Further Selective Schools Scholarship Tests Multiple Choice Yrs 5–8	Peter Howard
9781862941359	125	Year/Grade 4 Tests (O.C.) multiple choice Yrs 4–5	Peter Howard
9781862940598	64	Selective Schools Scholarship Tests Yrs 5–8	Peter Howard
9781862940758	73	Basic Skills Test Yrs 3–8 - Suitable preparation for NAPLAN Tests	Peter Howard
9781862941403	144	Year 3 Basic Skills Test - Suitable preparation for NAPLAN Tests	Peter Howard
9781862941632	153	Year 5 Basic Skills Test - Suitable preparation for NAPLAN Tests	Peter Howard
NAPLAN* Format Practice Tests			
9781921565434	235	Language Conventions Year 3 - NAPLAN* Format Practice Tests	Don Robens
9781921565441	236	Language Conventions Year 5 - NAPLAN* Format Practice Tests	Don Robens
9781921565533	245	Language Conventions Year 7 - NAPLAN* Format Practice Tests	Don Robens
9781921565458	237	Writing Year 3 - NAPLAN* Format Practice Tests	Alfred Fletcher
9781921565465	238	Writing Year 5 - NAPLAN* Format Practice Tests	Alfred Fletcher
9781921565519	243	Writing Year 7 - NAPLAN* Format Practice Tests	Alfred Fletcher
9781921565526	244	Writing Year 9 - NAPLAN* Format Practice Tests	Alfred Fletcher
9781921565472	239	Reading Year 3 - NAPLAN* Format Practice Tests	Alfred Fletcher
9781921565489	240	Reading Year 5 - NAPLAN* Format Practice Tests	Alfred Fletcher
9781921565557	247	Reading Year 7- NAPLAN* Format Practice Tests	Alfred Fletcher
9781921565496	241	Numeracy Year 3 - NAPLAN* Format Practice Tests	Don Robens
9781921565502	242	Numeracy Year 5 - NAPLAN* Format Practice Tests	Don Robens
9781921565540	246	Numeracy Year 7 - NAPLAN* Format Practice Tests	Don Robens
Basic Skills General Ability			
9781862942868	176	Giant Book of General Ability Tests Yrs 5–8	Graeme Wilson
9781862940857	81	General Aptitude Tests for Selective Yrs 5–8	Peter Howard
9781862941724	154	IQ Examples Level 1 Yrs 3–4	Peter Howard
9781862941731	155	IQ Examples Level 2 Yrs 5–8	Peter Howard
9781862940727	70	Learn to Think Level 1 Yrs 2–7	Peter Howard
9781862940734	71	Learn to Think Level 2 Yrs 5–8	Peter Howard
Basic Skills Language and Maths			
9781862940635	50	Kindergarten	Peter Howard
9781862940475	51	Year 1 Language/Mathematics	Peter Howard
9781862940482	52	Year 2 Language/Mathematics	Peter Howard
9781862940499	53	Year 3 Language/Mathematics	Peter Howard
9781862940505	54	Year 4 Language/Mathematics	Peter Howard
9781862940512	55	Year 5 Language/Mathematics	Peter Howard
9781862940529	56	Year 6 Language/Mathematics	Peter Howard
9781862940765	74	Year 7 Language/Mathematics	Peter Howard
9781862940864	82	Year 8 Language/Mathematics	Peter Howard
9781862941274	117	Maths & Language Problems Level 1 Yrs 1–2	Peter Howard
9781862941281	118	Maths & Language Problems Level 2 Yrs 5–8	Peter Howard
9781862941298	119	Maths & Language Problems Level 1 Yrs 1–2	Peter Howard
9781862941564	146	Challenging Maths Problems Yrs 5–8	
Basic Skills Maths			
9781862940642	65	Basic Skills Maths Level 1 Yrs 1–2	Peter Howard
9781862940659	66	Basic Skills Maths Level 2 Yrs 3–4	Peter Howard
9781862940666	67	Basic Skills Maths Level 3 Yrs 5–8	Peter Howard
9781862940994	95	Practice & Improve Your Maths Yr K–1	Arthur Baillie
9781862941007	96	Practice & Improve Your Maths Yr 2	Arthur Baillie
9781862941014	97	Practice & Improve Your Maths Yr 3	Arthur Baillie
9781862941021	98	Practice & Improve Your Maths Yr 4	Arthur Baillie
9781862941038	99	Practice & Improve Your Maths Yr 5	Arthur Baillie
9781862941045	100	Practice & Improve Your Maths Yr 6	Arthur Baillie
Easy Learn Maths			
9781862942844	145A	Basic Skills Easy– Learn Maths Pre/Kinder A	Valerie Marett
9781862942851	145B	Basic Skills Easy– Learn Maths Pre/Kinder B	Valerie Marett
9781862941410	130	Basic Skills Easy– Learn Maths 1A	Julin Tan
9781862941427	131	Basic Skills Easy– Learn Maths 1B	Julin Tan
9781862941434	132	Basic Skills Easy– Learn Maths 2A	Julin Tan

Coroneos Publications – Titles by Topic

ISBN	No.	Title	Author
9781862941441	133	Basic Skills Easy– Learn Maths 2B	Julin Tan
9781862941458	134	Basic Skills Easy– Learn Maths 3A	Julin Tan
9781862941465	135	Basic Skills Easy– Learn Maths 3B	Julin Tan
9781862941472	136	Basic Skills Easy– Learn Maths 4A	Julin Tan
9781862941489	137	Basic Skills Easy– Learn Maths 4B	Julin Tan
9781862941496	138	Basic Skills Easy– Learn Maths 5A	Julin Tan
9781862941502	139	Basic Skills Easy– Learn Maths 5B	Julin Tan
9781862941519	140	Basic Skills Easy– Learn Maths 6A	Julin Tan
9781862941526	141	Basic Skills Easy– Learn Maths 6B	Julin Tan
9781862941533	142	Basic Skills Easy– Learn Maths 7A Years 6–8	Valerie Marett
9781862941540	143	Basic Skills Easy– Learn Maths 7B Years 6–8	Valerie Marett
Basic Skills Science			
9781862941656	159	Succeeding in Science Yr 1	Graeme Wilson
9781862941663	160	Succeeding in Science Yr 2	Graeme Wilson
9781862941670	161	Succeeding in Science Yr 3	Graeme Wilson
9781862941687	162	Succeeding in Science Yr 4	Graeme Wilson
9781862941694	163	Succeeding in Science Yr 5	Graeme Wilson
9781862941700	164	Succeeding in Science Yr 6	Graeme Wilson
9781862941717	165	Succeeding in Science Yr 7	Graeme Wilson
9781862941809	170	Science and Technology Tests Book 1 Years 3–4	Graeme Wilson
9781862941816	171	Science and Technology Tests Book 2 Years 5–8	Graeme Wilson
Basic Skills Reading Comprehension			
9781862941823	173	Basic Skills – Giant Book of Reading/Comprehension Tests Yrs 5–8	Graeme Wilson
9781862940741	72	Reading/Comprehension Tests for Selective Schools Scholarship Exams Yrs 5–8	Peter Howard
9781862940536	57	Reading/Comprehension Level 1 Yrs 1–2	Peter Howard
9781862940543	58	Reading/Comprehension Level 2 Yrs 3–4	Peter Howard
9781862940550	59	Reading/Comprehension Level 3 Yrs 5–8	Peter Howard
9781862940925	88	First Comprehension Yrs K–3	Peter Howard
9781862941083	103	"Treasure Island" by Robert Louis Stevenson Yrs 3–8	Peter Howard
9781862941090	104	"Black Beauty" by Anna Sewell Yrs 3–8	Peter Howard
9781862941106	105	"Aladdin and "Ali Baba" Yrs K–3	Peter Howard
9781862941182	108	"Wind in the Willows" by Kenneth Grahame Yrs 3–8	Peter Howard
9781862941199	109	Advanced Reading/Comprehension Tests Yrs 7–8	Peter Howard
9781862941267	116	"The Loaded Dog" and other stories by Henry Lawson Yrs 5–8	Peter Howard
9781862941571	147	Advanced Reading/Comprehension Tests Yrs 8–10	Peter Howard
9781862941595	149	"Dot and the Kangaroo" Yrs 3–6	Peter Howard
9781862941625	152	"Aesop's Fables" Years 1–3	Peter Howard
9781862941311	121	"Dad & Dave" by Steele Rudd Yrs 3–8	Peter Howard
Basic Skills Spelling and Vocabulary			
9781862940567	60	Phonic/Reading Spelling Yrs K–3	Peter Howard
9781862940574	61	Spelling/Vocabulary Level 1 Yrs 1–2	Peter Howard
9781862940581	62	Spelling/Vocabulary Level 2 Yrs 3–4	Peter Howard
9781862940604	63	Spelling/Vocabulary Level 3 Yrs 5–8	Peter Howard
9781862941601	150	Vocabulary Builder Level 1 Years 3–4	Peter Howard
9781862941618	151	Vocabulary Builder Level 2 Years 5–8	Peter Howard
Basic Skills English, Writing, Language & Grammar			
9781862940963	92	English Level 1 Yrs 1–2	Peter Howard
9781862940970	93	English Level 2 Yrs 3–4	Peter Howard
9781862940987	94	English Level 3 Yrs 5–8	Peter Howard
9781862940895	85	Grammar & Punctuation Level 1 Yrs 3–4	Peter Howard
9781862940901	86	Grammar & Punctuation Level 2 Yrs 5–8	Peter Howard
9781862941205	110	Advanced Language Tests Yrs 8–10	Peter Howard
9781862941328	122	Advanced English Yrs 8–10	Peter Howard
9781862940710	69	Creative Writing Yrs 3–8	Peter Howard
9781862941366	126	Further creative Writing Yrs 5–8	Peter Howard
9781862941588	148	Advanced Creative Writing Yrs 8–10	Peter Howard
9781862940918	87	Written Expression for Selective Schools/Scholarship Exams Yrs 5–8	Peter Howard
Basic Skills HSIE / SOSE			
9781862940956	91	Activities in Social Studies Yrs 2–8	Peter Howard
9781862941397	129	All About Australia (a New Junior Geography)	Peter Howard
Basic Skills Fun Learning			
9781862941748	156	Fun Learning Activity Book 1 Ages 3–5	Peter Howard
9781862941755	157	Fun Learning Activity Book 2 Ages 6–8	Peter Howard

Coroneos Publications – Titles by Topic

9781862941762	158	Fun Learning Activity Book3 Ages 9–11	Peter Howard
9781862941335	123	Oz Slang All Ages	Peter Howard
9781862942141	204	More Oz Slang	Peter Howard
Basic Skills Early Years			
9781862941342	124	Introductory Comprehension Yrs K–3	Peter Howard
9781862941373	127	Pre-School Activities 1	Peter Howard
9781862941380	128	Pre-School Activities 2	Peter Howard
9781862940703	68	Reading Practice Yrs 1–3	Peter Howard
9781862940772	75	My First Reading 1000 Reading/Spelling Words yrs K–3	Peter Howard
9781862940819	77	Fun with Words Book 1 (colour) Yrs K–3	Peter Howard
9781862940826	78	Fun with Words Book 2 (colour) Yrs K–3	Peter Howard
9781862940833	79	Counting Practice Yrs K and Under	Peter Howard
9781862940840	80	First Reading Yrs k and Under	Peter Howard
9781862940932	89	First Creative Writing Yrs K–3	Peter Howard
9781862940949	90	Picture words Yrs K and under	Peter Howard
9781862941212	111	First Phonics Yrs K–3	Peter Howard
9781862941229	113	First phonic reading Yrs K–3	Peter Howard
9781862941236	114	Introductory Creative Writing Yrs K–3	Peter Howard
9781862941250	115	First Words Yrs K–3	Peter Howard
9781862941304	120	Second words Yrs K–3	Peter Howard
9781862940871	83	First handwriting Yrs K–2	Peter Howard
9781862940888	84	Handwriting Practice yrs 3–4	Peter Howard
Early Basic Skills NSW font			
9781921565564	266	Early Basic Skills Single Sounds – using NSW font	D.J. Ferguson
9781921565571	267	Early Basic Skills Simple Words & Sentences – using NSW font	D.J. Ferguson
9781921565588	268	Early Basic Skills Blending Consonants – using NSW font	D.J. Ferguson
9781921565595	269	Early Basic Skills Using Diagraphs – using NSW font	D.J. Ferguson
9781921565601	270	Early Basic Skills Written Text – Punctuation & Grammar – using NSW font	D.J. Ferguson
Early Basic Skills Victorian font			
9781921565618	366	Early Basic Skills Single Sounds – using Victorian font	D.J. Ferguson
9781921565625	367	Early Basic Skills Simple Words & Sentences – using Victorian font	D.J. Ferguson
9781921565632	368	Early Basic Skills Blending Consonants – using Victorian font	D.J. Ferguson
9781921565649	369	Early Basic Skills Using Diagraphs – using Victorian font	D.J. Ferguson
9781921565656	370	Early Basic Skills Written Text – Punctuation & Grammar – using Victorian font	D.J. Ferguson
Early Learning Skills (old editions)			
9781862941779	166	Early Learning Skills Book 1 Pre School/Kinder - Single Sounds	Denise Tyras
9781862941786	167	Early Learning Skills Book 2 Years K–1 - Three Letter Words	Denise Tyras
9781862941793	168	Early Learning Skills Book 3 Years 1–2 - Blends	Denise Tyras
9781862941854	172	Grammar, Punctuation and Vocabulary Years 1/2	Denise Tyras

DON ROBENS TITLES

Spelling			
9781862941946	188	Spelling Year 1	Don Robens
9781862941953	189	Spelling Year 2	Don Robens
9781862941960	190	Spelling Year 3	Don Robens
9781862941977	191	Spelling Year 4	Don Robens
9781862941984	192	Spelling Year 5	Don Robens
9781862941991	193	Spelling Year 6	Don Robens
Creative Writing			
9781862942028	196	Writing in Text Types Year 3	Don Robens
9781862942035	197	Writing in Text Types Year 4	Don Robens
9781862942042	198	Writing in Text Types Year 5	Don Robens
9781862942059	199	Writing in Text Types Year 6	Don Robens
Comprehension			
9781921565755	248	Comprehension Year 3	Don Robens
9781921565762	249	Comprehension Year 4	Don Robens
9781921565779	250	Comprehension Year 5	Don Robens
9781921565786	251	Comprehension Year 6	Don Robens

Coroneos Publications – Titles by Topic

AUSTRALIAN HOMESCHOOLING

Mathematics

9781862942158	501	Learning Multiplication 1 – Australian Homeschooling	Carmel Musumeci
9781862942165	502	Learning Multiplication 2 – Australian Homeschooling	Carmel Musumeci
9781921565298	531	Practise Your Addition and Subtraction to 20	Marett & Musumeci

Social Studies

9781862942509	503	Succeeding in Social Studies K	Valerie Marett
9781862942516	504	Succeeding in Social Studies 1	Valerie Marett
9781862942523	505	Succeeding in Social Studies 2	Valerie Marett
9781862942530	506	Succeeding in Social Studies 3	Valerie Marett
9781862942547	507	Succeeding in Social Studies 4	Valerie Marett
9781862942554	508	Succeeding in Social Studies 5	Valerie Marett
9781862942561	509	Succeeding in Social Studies 6	Valerie Marett
9781921565212	523	Australian History 1901–1945	Valerie Marett
9781921565335	535	Australian Government	Valerie Marett

English

9781862942219	510	Successful English 1	Valerie Marett
9781862942226	511	Successful English 2	Valerie Marett
9781862942479	512	Successful English 3A	Valerie Marett
9781862942486	513	Successful English 3B	Valerie Marett
9781921565250	527	Successful English 4A	Valerie Marett
9781921565267	528	Successful English 4B	Valerie Marett
9781921565342	536	Successful English 5A	Valerie Marett
9781921565359	537	Successful English 5B	Valerie Marett

Phonics

9781921565229	524	Revise Your Phonics 1	Valerie Marett
9781921565236	525	Revise Your Phonics 2	Valerie Marett

Spelling

9781862942233	514	Successful Spelling K	Valerie Marett
9781862942240	515	Successful Spelling 1	Valerie Marett
9781862942257	516	Successful Spelling 2	Valerie Marett
9781862942493	517	Successful Spelling 3	Valerie Marett
9781921565304	532	Successful Spelling 4	Valerie Marett
9781921565663	538	Successful Spelling 5	Valerie Marett
9781862942837	177	Learn to Read, Write & Spell 1 – Yrs K–1	Valerie Marett
9781862942820	178	Learn to Read, Write & Spell 2 – Yrs K–1	Valerie Marett
9781862942769	179	Learn to Read, Write & Spell 3 – Yrs 1–3	Valerie Marett
9781862942776	180	Learn to Read, Write & Spell 4 – Yrs 1–3	Valerie Marett
9781862942783	181	Learn to Read, Write & Spell 5 – Yrs 1–4	Valerie Marett
9781862942790	182	Learn to Read, Write & Spell 6 – Yrs 3–8	Valerie Marett
9781862942806	183	Learn to Read, Write & Spell 7 – Yrs 5–10	Valerie Marett

Testing

9781862942172	518	Test your Maths K–1	Valerie Marett
9781862942189	519	Test Your Maths 2	Valerie Marett
9781921565243	526	Test Your Maths 3	Valerie Marett
9781921565328	534	Test Your Maths 4	Valerie Marett
9781921565687	540	Test Your Maths 5	Valerie Marett
9781862942196	520	Test your Spelling & English K–1	Valerie Marett
9781862942202	521	Test your Spelling & English 2	Valerie Marett
9781862942578	522	Test your Reading & Phonics K–3	Valerie Marett
9781921565274	529	Test Your Spelling Years 3 & 4	Valerie Marett
9781921565281	530	Test Your English 3	Valerie Marett
9781921565311	533	Test Your English 4	Valerie Marett
9781921565670	539	Test Your English 5	Valerie Marett

Science

9781921565694	541	Secondary Science 7A	Frank Marett
9781921565700	542	Secondary Science 7B	Frank Marett
9781921565717	543	Secondary Science 7C	Frank Marett
9781921565724	544	Secondary Science 7D	Frank Marett
9781921565731	545	Secondary Science 8A	Frank Marett
9781921565748	546	Secondary Science 8B	Frank Marett

Coroneos Publications – Titles by Topic

MATHEMATICS

Secondary Texts

9781862940017	2	Year 9,10 Preparing for Year 11 Mathematics	Jim Coroneos
9781862940062	7	Year 11, 3 Unit Mathematics Course	Jim Coroneos
9781862940079	8	Year 12 3 Unit maths Course	Jim Coroneos
9781862940154	16	Complete 2 Unit Mathematics Course in 1 Volume	Jim Coroneos
9781862940178	18	Simplified 3 Unit Maths Course Vol 2	Jim Coroneos
9781862940185	19	Revised 4 unit Maths Course	Jim Coroneos
9781862940192	20	Supplement for 4 Unit Maths (large page)	Jim Coroneos

Past Papers and Solutions

9781921565380	35	HSC Mathematics Past Papers1990–2009 with Worked Solutions	Jim Coroneos & Others
9781921565373	36	HSC Extension 1 Mathematics Past Papers 1990–2009 with Worked Solutions	Jim Coroneos & Others
9781921565366	37	HSC Extension 2 Mathematics Past Papers 1990–2009 with Worked Solutions	Jim Coroneos & Others
9781921565397	175	HSC General Mathematics Past Papers 2001–2009 with Worked Solutions	Jim Coroneos & Others
9781921565427	206	HSC Mathematics 2001–2009 Past Papers & Worked Solutions	Jim Coroneos & Others
9781921565410	207	HSC Mathematics Extension 1 2001–2009 Past Papers & Worked Solutions	Jim Coroneos & Others
9781921565403	208	HSC Mathematics Extension 2 2001–2009 Past Papers & Worked Solutions	Jim Coroneos & Others

Past Papers and Solutions (Older Series)

9781862940321	33	Past HSC 2 Unit/3 Unit Maths Papers from 1967 with Answers	Jim Coroneos
9781862940338	34	Past HSC 3 Unit/4 Unit Maths Papers from 1967 with Answers	Jim Coroneos
9781862940673	47	Past NSW SC Maths Papers Advanced Level from 1983 with Answers/Solutions	Jim Coroneos
9781862940680	48	Past NSW SC Maths Papers Intermediate Level from 1983 with Answers/Solutions	Jim Coroneos
9781862941113	106	Past NSW SC Maths Paper General Level from 1988 with Answers/Solutions	Jim Coroneos

Practice (Specimen) Papers

9781862940260	27	Mathematics Practice papers with Answers	Jim Coroneos
9781862940277	28	Mathematics Extension 1 Practice Papers with Answers	Jim Coroneos
9781862940284	29	Mathematics Practice papers with Worked solutions	Jim Coroneos
9781862940291	30	Mathematics Extension 1 Practice Papers with Worked Solutions	Jim Coroneos
9781862940307	31	Mathematics Extension 1 Practice papers Plus Worked Solutions	Jim Coroneos
9781862940376	38	Short Question Ordinary/Credit Level Maths Papers with Answers	Jim Coroneos
9781862940383	39	Solutions to Short Question Ordinary Credit Level Maths Papers	Jim Coroneos
9781862940406	41	Short Question Advanced Level Maths Papers with Worked Solutions	Jim Coroneos
9781862940420	43	New Course Advanced Level Maths Papers with Answers	Jim Coroneos
9781862940437	44	New Course intermediate Level Maths Papers with Answers	Jim Coroneos
9781862940444	45	New Course Advanced Level Maths Papers with Worked Solutions	Jim Coroneos
9781862940451	46	New Course Intermediate Level Maths Papers with Worked Solutions	Jim Coroneos

Study Skills

9781862940000	1	How to Study Effectively Yrs 4–8	Coroneos & Smith

Coroneos Publications – Titles by Topic

EXCELLENCE SERIES

English

9781862942356	211	Excellence in English Year 1	Peter Howard
9781862942363	212	Excellence in English Year 2	Peter Howard
9781862942370	213	Excellence in English Year 3	Peter Howard
9781862942387	214	Excellence in English Year 4	Peter Howard
9781862942394	215	Excellence in English Year 5	Peter Howard
9781862942400	216	Excellence in English Year 6	Peter Howard
9781875695256		Excellence in English for Secondary Students	Peter Howard
9781875695331		English Skills in Use for Secondary Students	Peter Howard

Reading Skills

9781862942417	217	Excellence in Reading Skills Year 1	Peter Howard
9781862942424	218	Excellence in Reading Skills Year 2	Peter Howard
9781862942431	219	Excellence in Reading Skills Year 3	Peter Howard
9781862942448	220	Excellence in Reading Skills Year 4	Peter Howard
9781862942455	221	Excellence in Reading Skills Year 5	Peter Howard
9781862942462	222	Excellence in Reading Skills Year 6	Peter Howard
9781875695485		Practical Reading and Writing Skills	Peter Howard
9781875695324		Reading and Writing Skills for secondary students	Peter Howard
9781875695904		Excellence in Written Expression for secondary students	Peter Howard

Literacy

9781921565007	223	Excellence in Literacy Year 1	Peter Howard
9781921565014	224	Excellence in Literacy Year 2	Peter Howard
9781921565021	225	Excellence in Literacy Year 3	Peter Howard
9781921565038	226	Excellence in Literacy Year 4	Peter Howard
9781921565045	227	Excellence in Literacy Year 5	Peter Howard
9781921565052	228	Excellence in Literacy Year 6	Peter Howard
9781875695532		Practical Word Skills for Secondary Students	Peter Howard
9781875695652		Practical Spelling Skills	Judith Hall & Ron King
9781862942103		Middle School Vocabulary	Peter Howard
9781862942110		Dictionary Skills for Year 3/4	Peter Howard

Mathematics

9781921565069	229	Excellence in Maths Year 1	Peter Howard
9781921565076	230	Excellence in Maths Year 2	Peter Howard
9781921565083	231	Excellence in Maths Year 3	Peter Howard
9781921565090	232	Excellence in Maths Year 4	Peter Howard
9781921565106	233	Excellence in Maths Year 5	Peter Howard
9781921565113	234	Excellence in Maths Year 6	Peter Howard
9781875695386		Excellence in Maths 1 for secondary students	Fisher & Howard
9781875695416		Excellence in Maths 2 for secondary students	George Fisher

General Ability

9781862942066	209	Excellence in General Ability Year 4	Peter Howard
9781862942073	210	Excellence in General Ability Year 5	Peter Howard

PHONICS SERIES

9781862942080	261	Step by Step Phonics 1 NSW version *NSW Foundation Handwriting*	Peter Howard
9781862942097	262	Step by Step Phonics 2 NSW version *NSW Foundation Handwriting*	Peter Howard
9781921565199	263	Step by Step Phonic Reading 1 NSW version *NSW Foundation Handwriting*	Peter Howard
9781921565205	264	Step by Step Phonic Reading 2 NSW version *NSW Foundation Handwriting*	Peter Howard
9781862942127	361	Step by Step Phonics 1 Vic version *Victorian Modern Cursive*	Peter Howard
9781862942134	362	Step by Step Phonics 2 Vic version *Victorian Modern Cursive*	Peter Howard
9781862942332	363	Step by Step Phonic Reading 1 Vic version *Victorian Modern Cursive*	Peter Howard
9781862942349	364	Step by Step Phonic Reading 2 Vic version *Victorian Modern Cursive*	Peter Howard